Fluid Mechanics for Technicians 3/4

D.H.Bacon, M.Sc., C.Eng., M.I.Mech.E.

and

R.C.Stephens, M.Sc(Eng.), C. Eng., M.I.Mech.E.

First published 1982

British Library Cataloguing in Publication Data

Bacon, D. H.
Fluid mechanics for technicians 3/4.
1. Fluid mechanics
I. Title II. Stephens, R. C.
532'.002462 TA537

ISBN 0 408 01115 7

Printed and bound in Great Britain by
Mansell Bookbinders Ltd., Witham, Essex

PREFACE

This book is primarily intended to cover the Level III and Level IV units in Fluid Mechanics for the HTC and HTD courses; the text is largely based on the Authors' earlier work, *Mechanical Technology*. The book should also be suitable for degree and diploma students in other disciplines who require a fundamental outline of fluid mechanics.

The theory is presented concisely and is followed by worked examples which illustrate and amplify the theory. A large number of tutorial examples, with answers, are included in each chapter.

D. H. BACON
R. C. STEPHENS

CONTENTS

1 Introduction

Fluid Mechanics deals with fluids at rest (hydrostatics) and fluids in motion (hydrodynamics).

A fluid is defined as a substance which cannot resist shear stress and thus offers no resistance to change of shape. Fluids may be divided into liquids and gases; a quantity of liquid occupies a fixed volume, has a free horizontal surface and is virtually incompressible, whereas a quantity of gas completely fills the enclosing vessel and is easily compressed.

The majority of this work is concerned with liquids at rest or in motion and the most common liquid is, of course, water. The bulk modulus of water is about 2·1 GN/m^2, so that it compresses by approximately $0{\cdot}49 \times 10^{-3}$ of its volume for each N/m^2 of pressure. This is about seven times as compressible as steel but water may be assumed to be incompressible in most problems. The density of water is taken as 1 000 kg/m^3 (1 tonne/m^3 or 1 kg/litre) and the effects of temperature variation may be ignored.

When liquids are at rest, all behave in a similar way to water, the only significant variable being the density but for liquids in motion, viscosity also becomes an important factor. The viscosity of water is very small but for liquids such as oil, the viscosity has an important effect on the nature of the flow.

Problems concerning fluid flow involve the basic equations of mechanics, such as equilibrium of forces, Newton's laws, energy and momentum. These are used to determine pressure and velocity distribution, forces on stationary and moving bodies, flow rates in pipes and channels, viscous flow, power transmission, performance of pumps and turbines, etc.

An important technique in fluid mechanics is dimensional analysis and dynamic similarity, whereby the likely performance of large bodies such as ships and aeroplanes may be predicted from the performance of models. It is also useful in the analysis of pump and turbine performance and in other branches of engineering.

The flow of gases is more complex than the flow of liquids due to the effects of compressibility, density and temperature, which may be considerable. The general treatment of gases is beyond the scope of this book and gas flow problems are confined to those where variation of density may be ignored. Such problems may then be treated in the same way as liquid flow problems.

2 Pressure and pressure measurement

2.1 Units of pressure The basic unit of pressure is the N/m^2, which is termed the Pascal (Pa). The pressure exerted by the atmosphere is 101 325 N/m^2, so that the Pascal represents a very small pressure. More practical units are kN/m^2, MN/m^2 and the bar (1 bar = 10^5 N/m^2). It will be seen that atmospheric pressure is approximately 1 bar and is usually taken as such unless otherwise stated.

2.2 Pressure Since a fluid cannot resist shear stress, the pressure exerted by a fluid on any surface is always normal to that surface.

Consider a small triangular element ABC at a point in a fluid, Fig. 2.1. If the area of the face AC is a, then the area of AB is $a \sin\theta$ and that of BC is $a \cos\theta$. If the pressures on AC, AB and BC are p_1, p_2 and p_3 respectively, then, for horizontal equilibrium,

$$p_1 a \times \sin\theta = p_2 \times a \sin\theta, \qquad \text{i.e.} \quad p_1 = p_2$$

and for vertical equilibrium

$$p_1 a \times \cos\theta = p_3 \times a \cos\theta, \qquad \text{i.e.} \quad p_1 = p_3$$

Thus the pressure at a point in a fluid is the same in all directions.

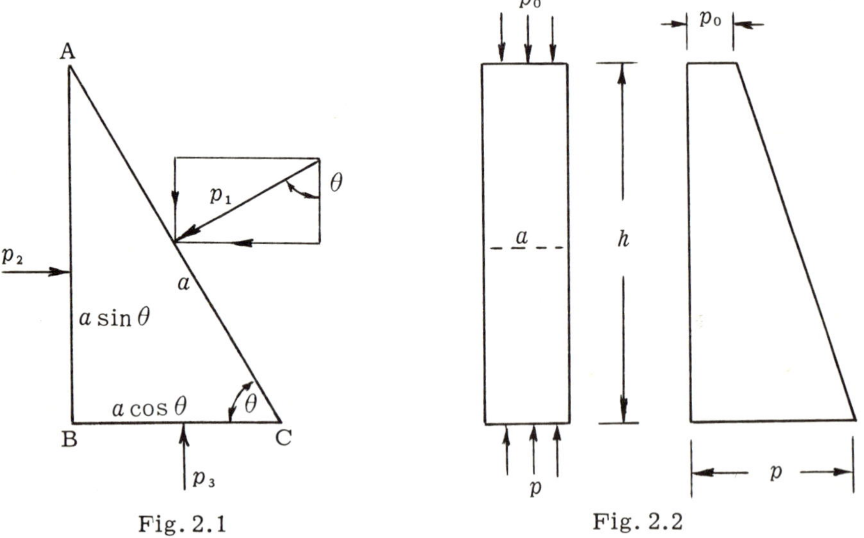

Fig. 2.1

Fig. 2.2

Fig. 2.2 shows a column of fluid at rest, the pressure on the upper and lower surfaces being p_0 and p respectively. If the density of the fluid is ρ, then for vertical equilibrium,

$$pa = p_0 a + \rho g a h$$

i.e.
$$p = p_0 + \rho g h \tag{2.1}$$

p is the total, or *absolute*, pressure and if p_0 represents atmospheric pressure, then $p - p_0$ is the *gauge* pressure.

Equation (2.1) shows that the pressure increases with depth but this variation is ignored in the case of gases due to the low densities involved. The only important exception is in the variation of pressure with altitude in the earth's atmosphere (Art. 3.6).

In the case of liquid pressure on a surface, it is only the gauge pressure which is relevant since the atmospheric pressure also acts on the reverse side of the surface. For such applications, the effective pressure at a depth h is merely that required to balance the weight of the liquid above,

i.e.
$$p = \rho g h \tag{2.2}$$

Equation (2.2) will give pressure in N/m^2 when ρ is in kg/m^3, g is in m/s^2 and h is in metres. It is often convenient to express pressure in terms of a head of liquid, given by $h = p/\rho g$. Thus an atmospheric pressure of 101·3 kN/m^2 is the pressure exerted by a column of water 10·34 m high or by a column of mercury (of specific gravity 13·6) 760 mm high.

Fig. 2.3 shows a solid body of cross-sectional area a immersed in a fluid. The horizontal forces on the body are in equilibrium but, in the vertical direction,

force on upper surface $= p_1 a = \rho g h_1 a$

and force on lower surface $= p_2 a = \rho g h_2 a$

$\therefore$ resultant upthrust $= \rho g a (h_2 - h_1)$

$=$ weight of displaced fluid

This is known as *Archimedes Principle.*

The upthrust acts through the centre of gravity of the displaced fluid.

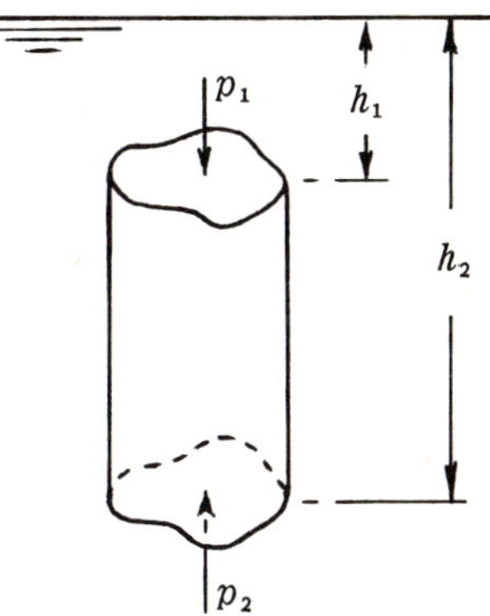

Fig. 2.3

2.3 Manometers A manometer is a simple form of pressure gauge which indicates the pressure in a pipe or vessel by the displacement of a liquid column. The following are the most common types of manometer and for simplicity, it will be assumed that the liquid whose pressure is being measured is water.

(*a*) *Piezometer tube, Fig. 2.4* This is simply a glass tube inserted in the top of the pipe ; the pressure then forces some water up the tube until equilibrium is attained.

The pressure is given by h m of water or $10^3 \times 9{\cdot}81h$ N/m^2.

(*b*) *U-tube, Fig. 2.5* If the pressure in the pipe is too high to use a piezometer tube, a U-tube filled with a heavy liquid, such as mercury, is used. If the specific gravity of the liquid is S, then x m of the liquid exerts the same pressure as Sx m of water.

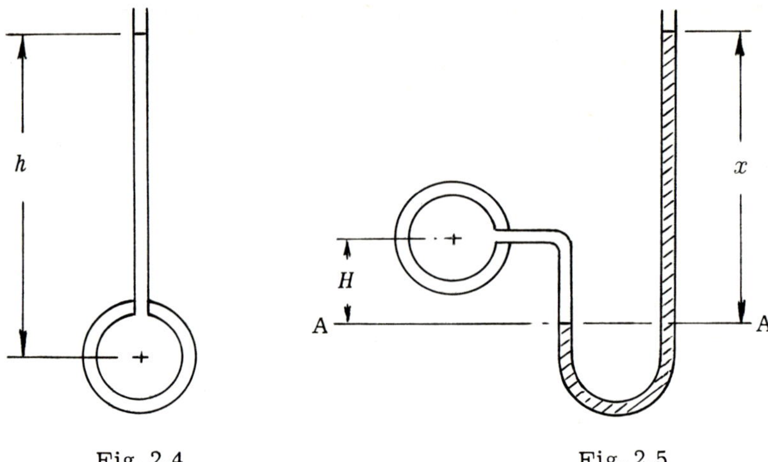

Fig. 2.4

Fig. 2.5

Hence, equating pressures on the two sides at level AA,

$$Sx = H + h$$ where h is the pressure in the pipe in metres of water.

i.e.

$$h = Sx - H \tag{2.3}$$

If a U-tube is used for measuring gas pressure, water or a lighter liquid may be used in the tube, depending upon the magnitude of the pressure being measured. In such an application the height H is irrelevant due to the low density of the gas.

(c) *U-tubes for pressure difference* To measure small differences in pressure between two points in a pipe, an inverted U-tube is used, Fig. 2.6.

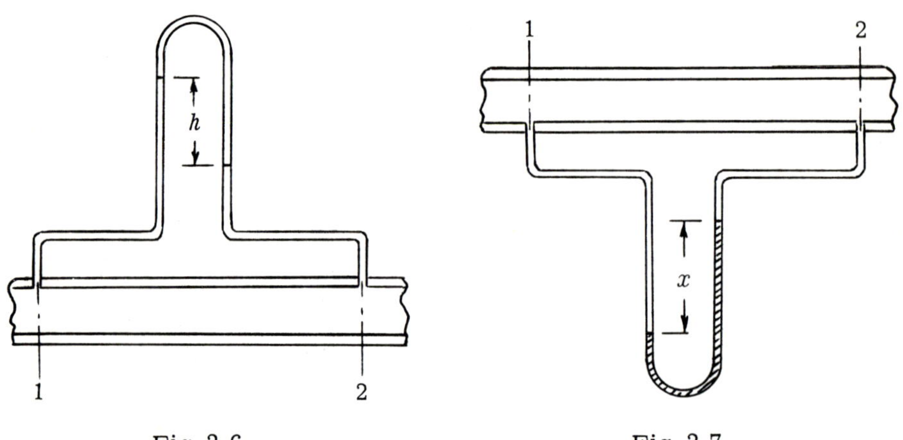

Fig. 2.6

Fig. 2.7

The difference in pressure between points 1 and 2 is given directly by

$$h_1 - h_2 = h \tag{2.4}$$

The sensitivity of such a manometer can be increased by filling the space above the water columns with a liquid lighter than water. If the specific gravity of this liquid is S, the difference of pressure is represented by x m of water, less Sx m of liquid.

i.e.
$$h_1 - h_2 = x(1 - S) \tag{2.5}$$

For the measurement of large pressure differences between two points the U-tube is arranged below the pipe and is filled with mercury or other heavy liquid, Fig. 2.7. If the specific gravity of the liquid is S then

$$h_1 - h_2 = (S - 1)x \tag{2.6}$$

(*d*) *General case of differential manometer* Let the density of the liquid in the pipe or vessel be ρ and that of the manometer liquid be ρ_f. If the difference between the levels of the measuring points is z and the difference in levels of the manometer liquid is x, Fig. 2.8, then, equating pressures at level XX,

$$p_1 + \rho g y = p_2 + \rho g(y + z - x) + \rho_f g x$$

i.e.
$$p_1 - p_2 = \rho g z + (\rho_f - \rho) g x$$

or
$$\frac{p_1 - p_2}{\rho g} = z + \left(\frac{\rho_f}{\rho} - 1\right) x \tag{2.7}$$

If the ends of the manometer are at the same level and the liquid in the pipe is water, this reduces to equation (2.6).

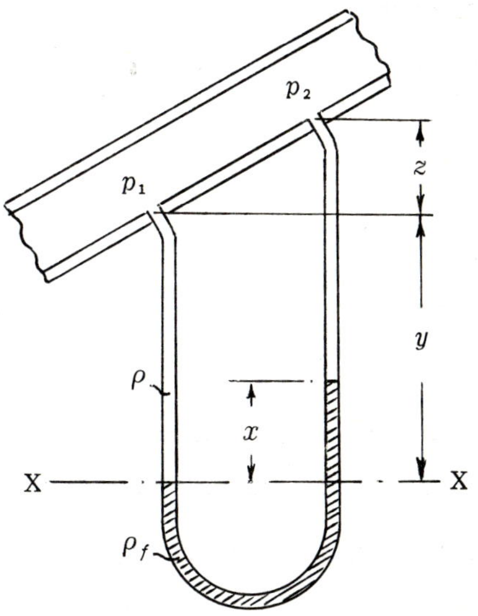

Fig. 2.8

This type of manometer may be used for a range of pressures by choosing the ratio ρ_f/ρ to give a reasonable value for x. It may also be modified by varying the cross-sectional area of the limbs (see example 2). In the application shown in Fig. 2.9, one limb has been made very large so that the liquid level remains virtually constant and only one limb reading need be taken. In Fig. 2.10, the reading limb is inclined to increase the sensitivity.

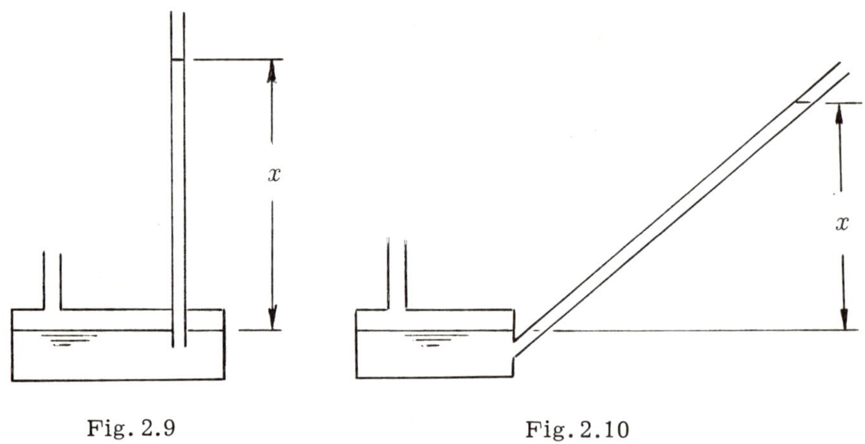

Fig. 2.9 Fig. 2.10

2.4 Bourdon gauge For pressure measurements where high precision is not required, a Bourdon gauge is commonly used, Fig. 2.11. The pressure is applied to a closed bent tube of elliptical cross-section which tends to straighten and this movement is used to move a needle round a dial through a toothed sector and pinion. The dial is normally calibrated to read zero at atmospheric pressure and hence indicates gauge pressure.

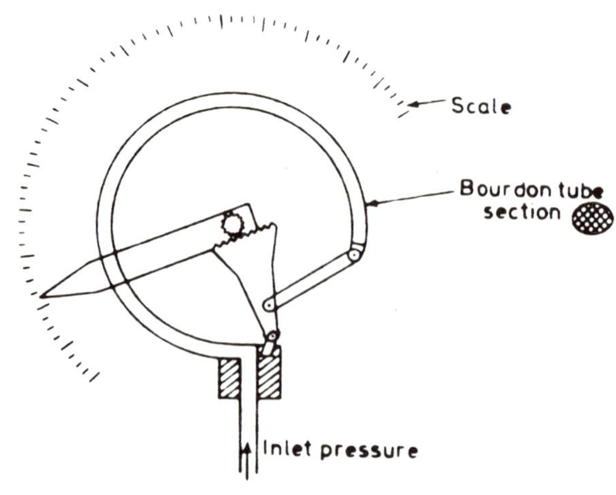

Fig. 2.11

2.5 Pressure transducers When pressure is changing, it is often necessary to record the pressure continuously over a period of time. A barograph is a simple device used to record atmospheric pressure but when changes are rapid, this simple bellows and mechanical linkage is inadequate and quick response devices are used which have an electrical output for recording on an oscilloscope, pen recorder of $u-v$ recorder or for feeding into a data logger, controller or microprocessor. A device which takes in a pressure signal and produces an electrical output signal proportional to the pressure is called a *transducer*.

In some pressure transducers, the fluid pressure acts on a diaphragm, to which is fitted a resistance strain gauge. Stretching of the diaphragm due to the pressure causes a change in the resistance of the strain gauge which, with the aid of a suitable circuit, produces an electric potential proportional to the pressure. The potential can then be applied to an oscilloscope, controller, etc., as shown in Fig. 2.12.

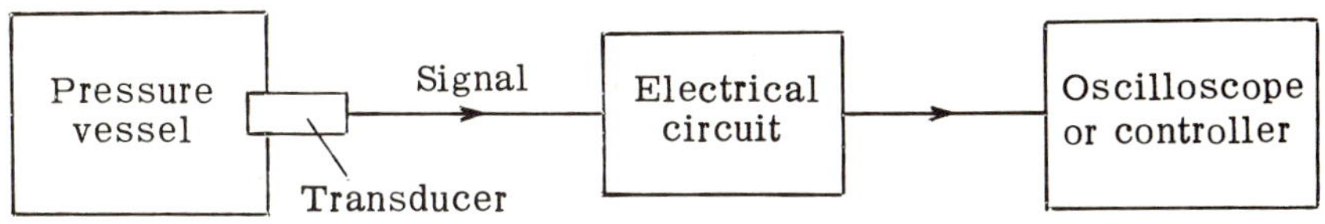

Fig. 2.12

Other transducers use change in capacitance or impedance of the sensing device or make use of a piezoelectric crystal to produce a small electrical signal under the action of the applied pressure.

2.6 Micromanometers For the measurement of very small pressure differences, a micromanometer is used. Two liquids are used and the ends are enlarged to give a large miniscus movement for a small pressure difference.

Fig. 2.13 shows such a manometer. The density of the fluid under investigation is ρ and those of the manometer liquids are ρ_1 and ρ_2. The cross-sectional areas of the tube and reservoir are a and A respectively.

Let the applied pressures be p_1 and p_2, where $p_1 > p_2$. If this results in a movement z in the reservoir levels and a difference of levels in the tube x, then

$$a \times \frac{x}{2} = A \times z$$

or

$$z = \frac{ax}{2A} \tag{2.8}$$

If the initial level of the common surface below the reservoir levels is y, then, equating pressures at the new common surface at level XX,

$$p_1 + \rho g(h+z) + \rho_1 g\left(y + \frac{x}{2} - z\right) = p_2 + \rho g(h-z) + \rho_1 g\left(y - \frac{x}{2} + z\right) + \rho_2 g x$$

which simplifies to
$$p_1 - p_2 = g\{(\rho_2 - \rho_1)x + 2(\rho_1 - \rho)z\}$$

$$= gx\left\{(\rho_2 - \rho_1) + (\rho_1 - \rho)\frac{a}{A}\right\} \tag{2.9}$$

substituting for z from equation (2.8).

If the ratio $\frac{a}{A}$ is very small, this reduces to

$$p_1 - p_2 = gx(\rho_2 - \rho_1) \tag{2.10}$$

Thus, if the difference of the manometer liquid densities, $\rho_2 - \rho_1$, is small, the reading x is large for a small applied pressure difference, $p_1 - p_2$.

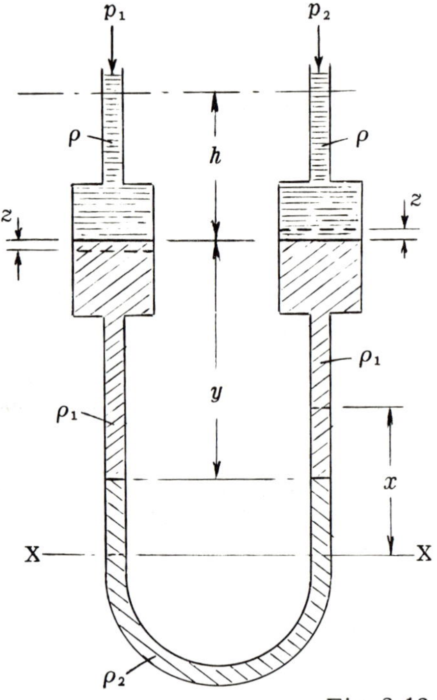

Fig. 2.13

To improve accuracy in reading this type of instrument, a graduated telescope may be used to view the meniscus and the whole instrument may be fitted with a micrometer screw to return the meniscus to its original position. The pressure difference is then obtained from the micrometer movement. Examples of these specialist instruments are the Chattock gauge and the Krell gauge, which can measure pressure differences as small as 0·02 N/m² (0·002 mm H_2O).

1. *A cylindrical buoy floats in sea water with its axis vertical so that it is two-thirds submerged. The buoy is 0·8 m in diameter and 2 m in height and is fabricated from iron plate 12 mm thick. Calculate the mass of iron chain securing the buoy.*

The density of iron is 7 700 kg/m³ and of sea water is 1 030 kg/m³.

Fig. 2.14 shows the buoy in equilibrium under the action of the weight W, the upthrust U and the chain force F.

The upthrust is equal to the weight of water displaced,

i.e. $$U = \frac{\pi}{4} \times 0{\cdot}8^2 \times \frac{2}{3} \times 2 \times 1\,030 \times 9{\cdot}81 = 6\,750 \text{ N}$$

Weight of buoy,

$$W = \left(\pi \times 0{\cdot}8 \times 2 + 2\times\frac{\pi}{4}\times 0{\cdot}8^2\right)\times\frac{12}{10^3}\times 7\,700\times 9{\cdot}81$$

$$= 5\,450 \text{ N}$$

$$\therefore\ F = U - W$$

$$= 6\,750 - 5\,450 = 1\,300 \text{ N}$$

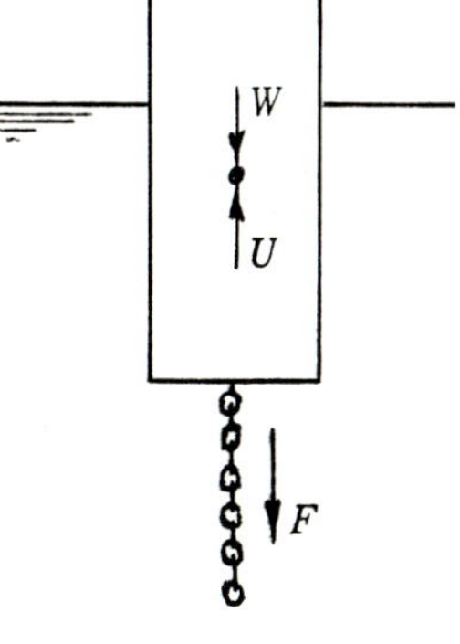

Fig. 2.14

The chain force is the difference between the weight of the chain and the upthrust on the chain. If the volume of the chain is V, then

$$1\,300 = 7\,700\times 9{\cdot}81\times V - 1\,030\times 9{\cdot}81\times V$$

from which $V = 0{\cdot}0199 \text{ m}^3$

Therefore mass of chain $= 7\,700\times 0{\cdot}0199 = \underline{152 \text{ kg}}$

2. *A pressure gauge for the measurement of small pressures consists of a vertical glass U-tube with internal diameter 10 mm. The upper ends of the tube are enlarged to form reservoirs of diameter 60 mm. One limb contains water and is connected to the unknown pressure while the other contains paraffin and is open to the atmosphere. Calculate the gauge pressure applied to the water limb when the water-paraffin meniscus is displaced 10 mm.*

Density of paraffin = 850 kg/m³.

Let the initial level of the common surface be at XX, Fig. 2.15, and let the heights of the paraffin and water columns above this level be h_p and h_w respectively.

Equating pressures at XX :–

$$\rho_p g h_p = \rho_w g h_w$$

$$\therefore\ h_w = \frac{850}{1\,000} h_p$$

$$= 0{\cdot}85\, h_p$$

When the common surface is displaced 10 mm,

$$z = 10 \times \frac{\frac{\pi}{4}\times 10^2}{\frac{\pi}{4}\times 60^2}$$

$$= \frac{10}{36} \text{ mm}$$

$$= 0{\cdot}000\,278 \text{ m}$$

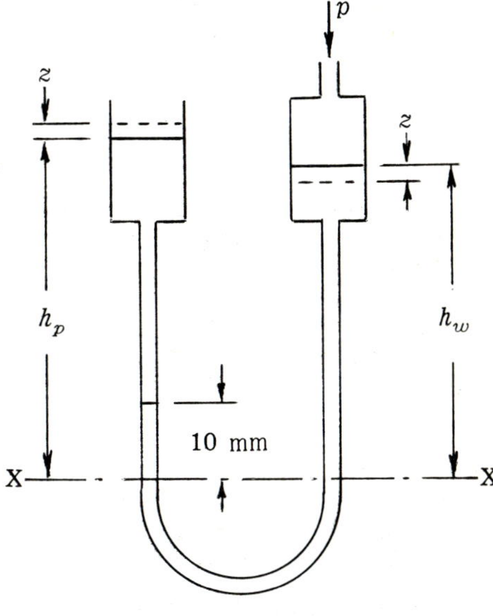

Fig. 2.15

Equating gauge pressures at level XX when the pressure p is applied to the water limb,

$$\rho_p g (h_p + z - 0{\cdot}01) + \rho_w g \times 0{\cdot}01 = \rho_w g (h_w - z) + p$$

i.e.

$$850g (h_p + 0{\cdot}000\,278 - 0{\cdot}01) + 1\,000g \times 0{\cdot}01 = 1\,000g (0{\cdot}85h_p - 0{\cdot}000\,278) + p$$

from which $$p = \underline{19{\cdot}77 \text{ N/m}^2}$$

This is a pressure of approximately 0·000 2 atmospheres.

3. *An open vertical cylinder has a connection from the bottom to one arm of a vertical U-tube containing mercury, the other arm being open to atmosphere. When the cylinder is empty, the level of the mercury in each of the tubes is initially 2 m below a fixed point A on the cylinder. Water is now poured into the cylinder up to the level of A. Determine the difference of levels of mercury in the two tubes.*

The cylinder diameter is 24 mm and the diameter of the tube is 6 mm. If a frictionless plunger weighing 40 N is now inserted into the cylinder, determine the new difference in mercury levels, assuming no leakage.

Relative density of mercury = 13·6.

If the difference in levels of mercury is x, Fig. 2.16, then, equating pressures at level XX,

$$2 + \frac{x}{2} = 13{\cdot}6x$$

$$\therefore \; x = \underline{0{\cdot}152\,7 \text{ m}}$$

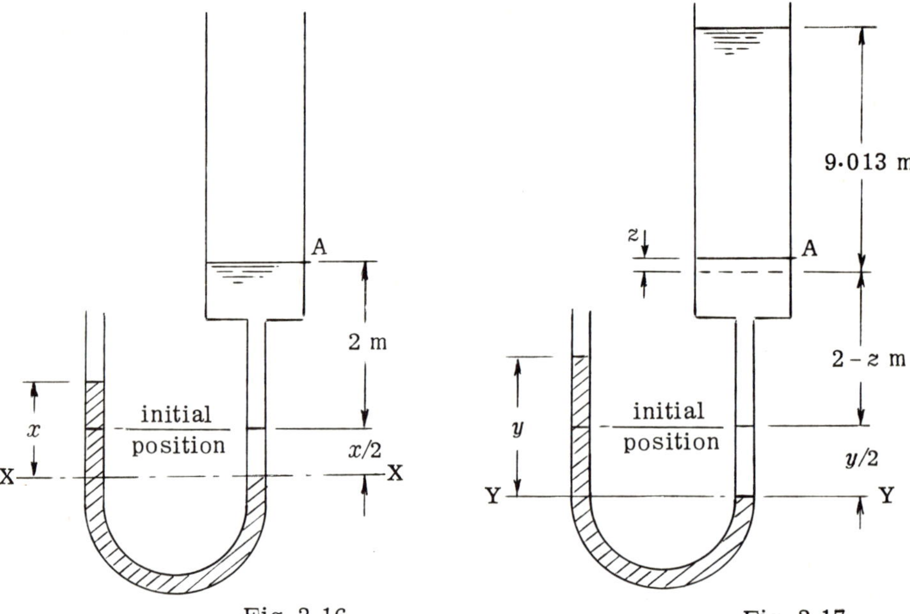

Fig. 2.16

Fig. 2.17

When the plunger is added, let the new difference in levels be y, Fig. 2.17. Then surface in cylinder falls a distance z such that

$$\frac{\pi}{4}\times 0{\cdot}024^2\times z = \frac{\pi}{4}\times 0{\cdot}006^2\times\left(\frac{y}{2}-\frac{x}{2}\right)$$

from which $$z = \frac{y - 0{\cdot}1527}{32}$$

The effect of the weight of the plunger is equivalent to adding further water of the same weight. Hence, if the additional depth of water is h,

$$\frac{\pi}{4}\times 0{\cdot}024^2\times h\times 10^3\times 9{\cdot}81 = 40$$

from which $$h = 9{\cdot}013 \text{ m}$$

Equating pressures at level YY,

$$9{\cdot}013 + (2 - z) + \frac{y}{2} = 13{\cdot}6y$$

i.e. $$11{\cdot}013 - \frac{y - 0{\cdot}1527}{32} = 13{\cdot}1y$$

from which $$y = \underline{0{\cdot}839 \text{ m}}$$

NOTE If the small movement z is ignored, $y = \underline{0{\cdot}841 \text{ m}}$

4. In order to lay out an anchor of mass 600 kg, it is suspended below a boat by a wire strop, as shown in Fig. 2.18. The density of the anchor material is 7 000 kg/m³. Determine the tensions in the strop at points A and B. The density of sea water is 1 030 kg/m³. (*Ans.*: 5·015 kN; 3·54 kN)

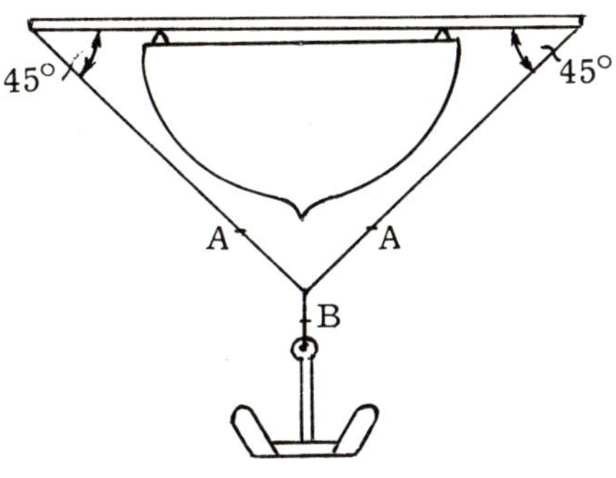

Fig. 2.18

5. A small sunken vessel is to be raised by filling the hold with watertight drums of 5 m³ capacity. The drums each have a mass of 120 kg and the vessel commences to rise when 70 of the drums have been pumped out. The total number of drums in the hold is 100. Calculate the mass of the vessel, neglecting the volume of metal in the vessel and drums. The density of sea water is 1025 kg/m³. (*Ans.*: 346·5 t)

6. A mercury U-tube manometer is used to measure the difference in pressure across a flowmeter in a horizontal pipe, the water from the pipe being in contact with the mercury in the manometer. The meter determines the flow rate of water according to the relation $Q = 0{\cdot}26\sqrt{h}$ when Q is the flow rate in m^3/s and h is the difference in head across the meter in metres of water.

Calculate the manometer reading when the flow rate is 0·7 m^3/s. The relative density of mercury is 13·6. (*Ans*.: 0·573 m)

7. A differential pressure gauge for the measurement of small air pressures consists of a vertical glass U-tube having a uniform cross-section of internal diameter 6 mm. The upper ends of the tube are enlarged to a diameter of 48 mm. The gauge is filled with water and kerosene and the free surfaces of the liquids are in the enlarged ends. The air pressure to be measured is applied to the top of the limb containing water, the other limb being open to atmosphere. Calculate the gauge pressure of the air which will cause a displacement of the meniscus of 25 mm from the balance position. The relative density of kerosene is 0·8. (*Ans*.: 55·95 N/m^2)

8. Fig. 2.19 shows a special manometer for use with a flowmeter which determines the flow rate in a horizontal pipe according to the relation $Q = 0{\cdot}07\sqrt{h}$ where Q is the flow rate in m^3/s and h is the difference in head across the meter in metres of water. By how much will level A move when the flow rate is 0·08 m^3/s? The diameter of the larger limb is 3·8 times that of the smaller limb and the relative density of the mercury is 13·6. (*Ans*.: 97·1 m)

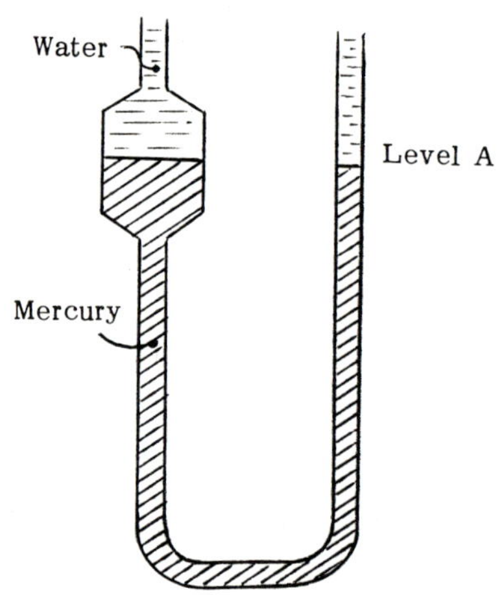

Fig. 2.19

9. A manometer of the type shown in Fig. 2.15 has a U-tube of 12 mm bore and enlarged ends of 60 mm diameter. The liquids used in the manometer are oil of density 890 kg/m^3 and water of density 1 000 kg/m^3. An air pressure 25 N/m^2 greater than atmospheric is applied to the water limb and the oil limb is left open to the atmosphere. Calculate the movement of the oil-water meniscus. (*Ans*.: 13·73 mm)

10. A manometer of the type shown in Fig. 2.13 has a U-tube of 12 mm bore and enlarged ends of 60 mm diameter. The liquids used in the manometer are oil of density 890 kg/m^3 and water of density 1 000 kg/m^3. A pressure difference of 25 kN/m^2 is applied by a gas of density 1·1 kg/m^3 to the liquid in the enlarged ends. Calculate the reading of the manometer. (*Ans*.: 17·4 mm)

3 Hydrostatics

3.1 Forces on immersed surfaces Since the pressure in a liquid increases with the depth, the resultant force on an immersed surface will depend on the average pressure and will always act at a point below the centroid of the area.

3.2 Total force on a vertical surface Fig. 3.1 shows a vertical rectangular surface of breadth b extending to a depth h below the surface. The pressure distribution diagram is a triangle, of maximum ordinate $\rho g h$.

The total force P is the average ordinate of the triangle (which is the pressure at the centroid G), multiplied by the area bh.

i.e. $$P = \tfrac{1}{2}\rho g h \times bh = \tfrac{1}{2}\rho g b h^2 \qquad (3.1)$$

This force acts through the centroid of the triangle, which is at a depth $2h/3$ below the surface. The point of application of the force, P, is called the *centre of pressure*.

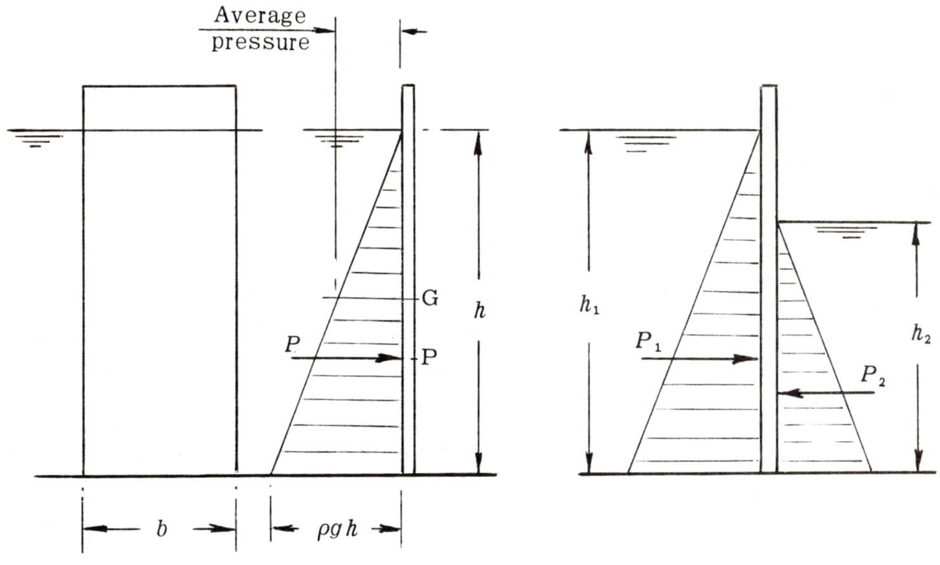

Fig. 3.1 Fig. 3.2

If a vertical surface has water on both sides, such as a lock gate, Fig. 3.2, then

$$P_1 = \tfrac{1}{2}\rho g b h_1^2 \qquad \text{acting at } 2h_1/3 \text{ below the surface}$$

and

$$P_2 = \tfrac{1}{2}\rho g b h_2^2 \qquad \text{acting at } 2h_2/3 \text{ below the surface}$$

The resultant force, $P = P_1 - P_2$ and the height h of its point of application from the base is obtained by taking moments about the base,

i.e.
$$Ph = P_1\frac{h_1}{3} - P_2\frac{h_2}{3}$$

As $h_2 \to h_1$, $P \to 0$ and $h \to \dfrac{h_1}{2}$.

3.3 Total force on any plane surface Fig. 3.3 shows a plane surface of area a immersed in a liquid, the surface being inclined at an angle θ to the horizontal.

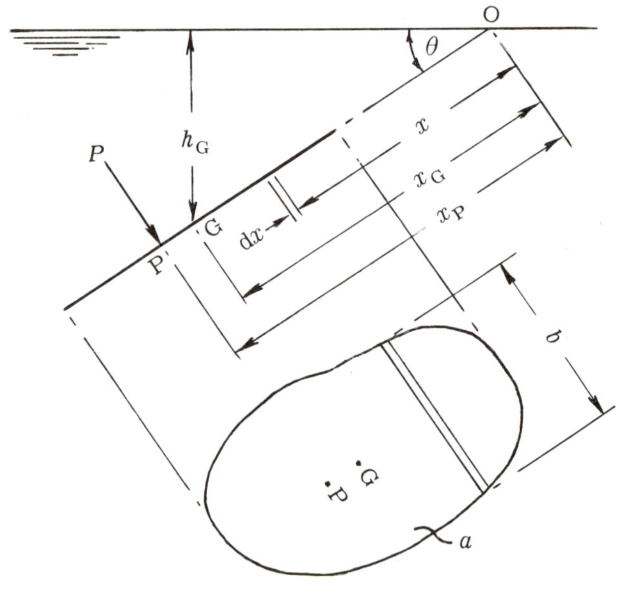

Fig. 3.3

Pressure on elementary strip at distance x from O

$$= \rho g x \sin\theta$$

$$\therefore \text{ force on strip} = \rho g x \sin\theta \times b\,dx$$

$$\therefore \text{ total force on surface} = \rho g \sin\theta \int xb\,dx$$

But $\int xb\,dx$ is the total first moment of the area a about O, which is equal to ax_G.

Thus $$P = \rho g a x_G \sin\theta = a(\rho g h_G)$$

$$= \text{area} \times \text{pressure at centroid} \qquad (3.2)$$

To find the line of action of P, moments are taken about the point O.

$$\text{Moment of force on strip about O} = \rho g x \sin\theta\, b\, dx \times x$$

$$\therefore \text{ total moment about O} = \rho g \sin\theta \int b x^2\, dx$$

But $\int b x^2\, dx$ is the second moment of the area about the point O, I_O

i.e. $$\text{total moment about O} = \rho g \sin\theta\, I_O$$

Also $$\text{moment of } P \text{ about O} = P x_P = \rho g a x_G \sin\theta \,.\, x_P$$

$$\therefore x_P = \frac{\rho g \sin\theta\, I_O}{\rho g \sin\theta\, a x_G} = \frac{I_O}{a x_G}$$

i.e. $$\text{distance of centre of pressure from O} = \frac{\text{second moment of area about O}}{\text{first moment of area about O}} \qquad (3.3)$$

By the theorem of parallel axes,

$$I_O = I_G + a x_G^2 = a(k_G^2 + x_G^2)$$

$$\therefore x_P = \frac{a(k_G^2 + x_G^2)}{a x_G} = x_G + \frac{k_G^2}{x_G}$$

Thus x_P is always greater than x_G but approaches x_G (i) if the distance x_G increases due to increased depth of immersion and (ii) if the inclination θ decreases.

For surfaces which can be divided into triangles, use can be made of the equimomental system *. For the vertical triangular surface shown in Fig. 3.4, the area a is replaced by areas $a/3$ imagined concentrated at the mid-points of the sides.

If the depths of these areas below the surface are α, β and γ, the depth of the centre of pressure h_P is given by

$$h_P = \frac{\frac{a}{3}\alpha^2 + \frac{a}{3}\beta^2 + \frac{a}{3}\gamma^2}{\frac{a}{3}\alpha + \frac{a}{3}\beta + \frac{a}{3}\gamma}$$

$$= \frac{\alpha^2 + \beta^2 + \gamma^2}{\alpha + \beta + \gamma}$$

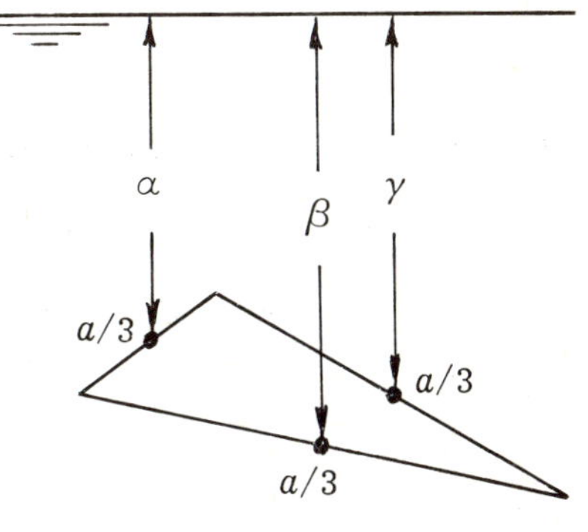

Fig. 3.4

* See *Mechanical Science for Higher Technicians*, Bacon and Stephens, Art. 3.1.

3.4 Forces on curved surfaces The force on a curved surface is obtained by considering the equilibrium of a wedge of fluid bounded by the curved surface and other suitably chosen plane surfaces.

For the curved surface of a dam, Fig. 3.5, the water contained within the wedge ABC is in equilibrium under its weight W, the force P exerted on the face AC by the adjacent water, and the reaction of the curved surface BC. The magnitude and direction of the reaction R is obtained from the triangle of forces and its position is determined by the concurrency of R with the forces P and W. The force exerted *on* the dam is then equal and opposite to that exerted *by* the dam on the wedge ABC.

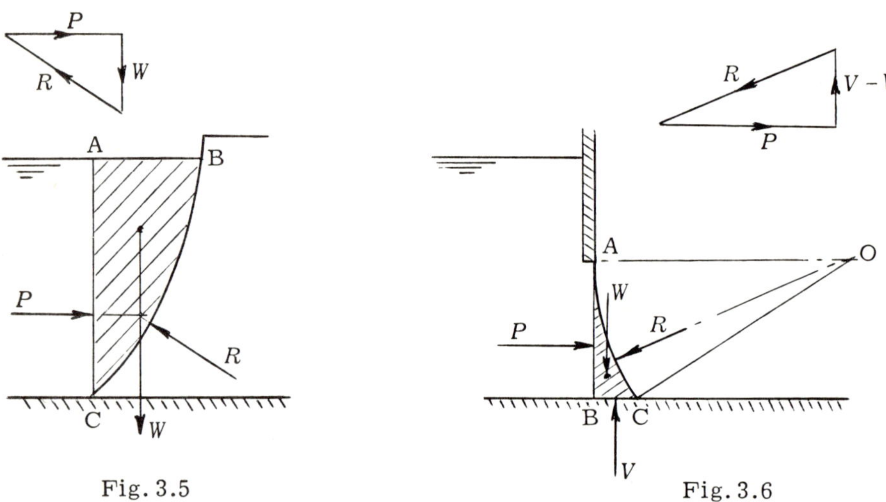

Fig. 3.5 Fig. 3.6

For a cylindrical or spherical surface, it is useful to note that, since the reaction R is normal to the surface, it must pass through the centre of curvature. Thus, for the sector gate shown in Fig. 3.6, the wedge ABC is in equilibrium under the horizontal force P exerted on the face AB by the adjacent water, the upward force $V - W$ where V is the vertical reaction of the base BC and W is the weight of water enclosed in the wedge and the reaction R. The magnitude and direction of R is determined by the triangle of forces as before but it is unnecessary to determine the position of the vertical force $V - W$ since R must pass through the point O.

If such a gate is hinged at O, no torque will be required to overcome the fluid force when the gate is opened or closed.

When a surface is curved in one plane only, as in the foregoing examples, it is usual to consider the forces acting on a unit length of surface.

For the hemispherical body shown in Fig. 3.7, the forces acting on it are the same as those acting on the liquid displaced, these being the horizontal force P acting on the face AB, the weight W of the liquid displaced and the reaction R on the curved surface, which passes through the centre O. It is again unnecessary to determine the position of the line of action of W.

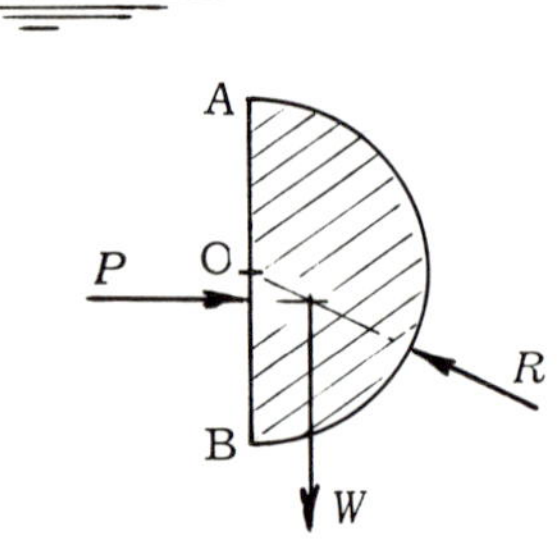

Fig. 3.7

3.5 Surfaces subjected to gas pressure The density of a gas is small and so the pressure exerted on a plane or curved surface by a gas is assumed to be uniform. Thus for a plane surface

$$\text{force} = \text{pressure} \times \text{area}$$

and this acts at the centroid of the area.

For a curved surface such as the hemispherical end of the gas cylinder shown in Fig. 3.8, the gas enclosed within the end is in equilibrium under the force on AB and the reaction of the curved surface,

i.e. $$R = p\pi r^2$$

A force equal and opposite to R, shown dotted, represents the force exerted *by* the gas *on* the surface.

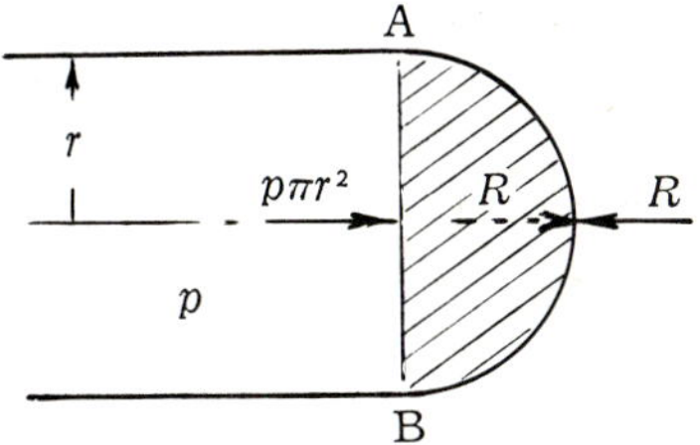

Fig. 3.8

3.6 The atmosphere Air may be regarded as a perfect gas and when calculating the effect of altitude on pressure, it is assumed that the temperature falls at the rate of 0·006 5°C/m (called the *lapse rate*). The relation between pressure and density is given by

$$p = c\rho^n$$

where c is a constant and n has the value 1·235.

Let the pressure at a height h above sea-level be p and let sea-level pressure, density and temperature be p_0, ρ_0 and T_0 respectively. If the pressure falls by dp as the height increases by dh, Fig. 3.9,

$$\mathrm{d}p = -\rho g\, \mathrm{d}h$$

or $$\mathrm{d}h = -\frac{\mathrm{d}p}{\rho g}$$

$$\frac{p}{\rho^n} = c \quad \text{so that} \quad \frac{1}{\rho} = \frac{c^{1/n}}{p^{1/n}}$$

Hence $$\mathrm{d}h = -\frac{c^{1/n}}{g}\left(\frac{\mathrm{d}p}{p^{1/n}}\right)$$

Integrating $$h = -\frac{c^{1/n}}{g}\left[\frac{p^{\frac{n-1}{n}}}{\frac{n-1}{n}}\right]_{p_0}^{p}$$

$$= \frac{n}{n-1}\frac{c^{1/n}}{g}\left[p_0^{\frac{n-1}{n}} - p^{\frac{n-1}{n}}\right]$$

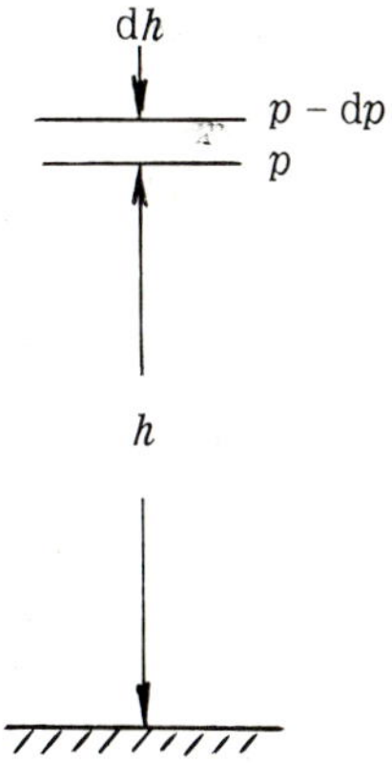

Fig. 3.9

But $$c^{1/n} = \frac{p^{1/n}}{\rho} = \frac{p_0^{1/n}}{\rho_0}$$

$$\therefore\ h = \frac{n}{n-1}\left[\frac{p_0}{\rho_0 g} - \frac{p}{\rho g}\right]$$

$$= \frac{n}{n-1}\frac{p_0}{\rho_0 g}\left[1 - \frac{p}{p_0}\cdot\frac{\rho_0}{\rho}\right]$$

But
$$\frac{\rho_0}{\rho} = \left(\frac{p_0}{p}\right)^{1/n}$$

$$\therefore\ h = \frac{n}{n-1}\frac{p_0}{\rho_0 g}\left[1 - \left(\frac{p}{p_0}\right)^{\frac{n-1}{n}}\right] \qquad (3.4)$$

For air, $p_0 = 101{\cdot}3\ \text{kN/m}^2$ and $\rho_0 = 1{\cdot}225\ \text{kg/m}^3$.

Hence, from equation (3.4),

$$h = \frac{1{\cdot}235}{1{\cdot}235 - 1} \times \frac{101{\cdot}3}{1{\cdot}225 \times 9{\cdot}81}\left[1 - \left(\frac{p}{101{\cdot}3}\right)^{\frac{1{\cdot}125-1}{1{\cdot}235}}\right]$$

from which
$$p = 101{\cdot}3\,(1 - 0{\cdot}000\,022\,6\,h)^{5{\cdot}255}\ \text{kN/m}^2 \qquad (3.5)$$

1. *A tank containing fuel of density 850 kg/m³ has an automatic device to prevent overflow; this consists of a rectangular door hinged at the upper edge and held closed by a spring, as shown in Fig. 3.10. The door is 0·3 m deep and 0·2 m wide and the spring exerts a force of 1 kN.*

Calculate the level of the fuel above the hinge when the door is about to open.

Let the surface of the fuel be a height h above the hinge.

From equation (3.2),

force on door due to oil pressure

= area × pressure at centroid

i.e.
$$P = (0{\cdot}2 \times 0{\cdot}3) \times (850 \times 9{\cdot}81\,[h + 0{\cdot}15])$$

$$= 500\,(h + 0{\cdot}15)\ \text{N}$$

From equation (3.3), depth of centre of pressure,

$$x_P = \frac{\text{second moment of area about surface}}{\text{first moment of area about surface}}$$

$$= \frac{\dfrac{bd^3}{12} + bd\,(h + 0{\cdot}15)^2}{bd\,(h + 0{\cdot}15)}$$

$$= \frac{0{\cdot}3^2}{12\,(h + 0{\cdot}15)} + (h + 0{\cdot}15)$$

$$= \frac{0{\cdot}0075}{h + 0{\cdot}15} + (h + 0{\cdot}15)$$

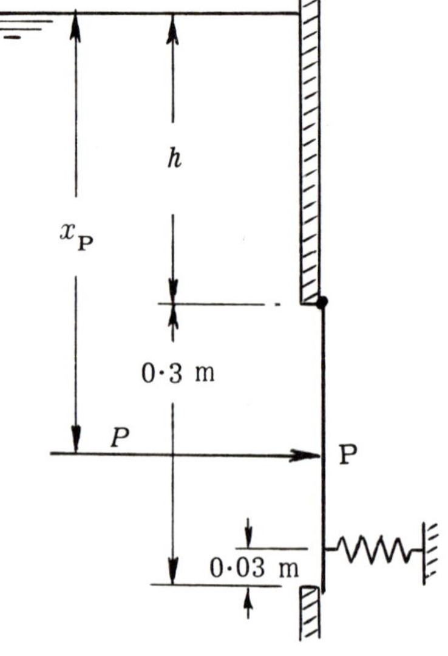

Fig. 3.10

The door will commence to open when the moment of the spring force about the hinge is equal to that of the water force about the hinge,

i.e. $$1 \times 10^3 (0{\cdot}3 - 0{\cdot}03) = 500(h + 0{\cdot}15)\left\{\frac{0{\cdot}0075}{h + 0{\cdot}15} + (h + 0{\cdot}15) - h\right\}$$

from which $$\underline{h = 3{\cdot}40 \text{ m}}$$

2. *Each gate of a lock is 6 m high and 2 m wide, and is supported on hinges, situated 0·6 m from the top and bottom. The angle between the gates when they are closed is 140°. If the depths of water on the two sides are 5·4 m and 1·5 m, find the magnitude and position of the resultant force on each gate, the reaction between the gates and the magnitude and direction of the reactions at the hinges.*

The forces acting on one gate are as shown in Fig. 3.11.

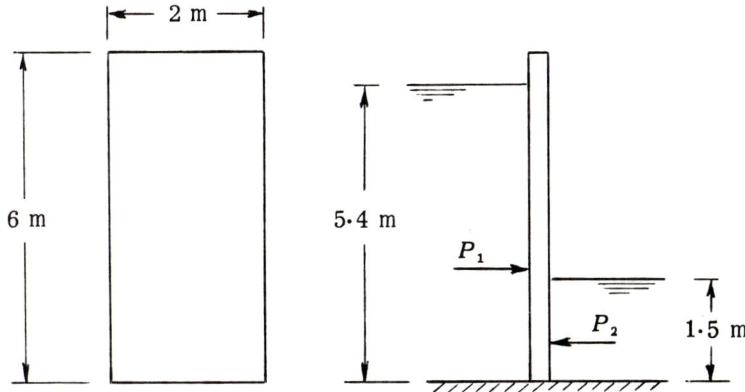

Fig. 3.11

$$P_1 = \tfrac{1}{2}\rho g b h^2 \quad . \quad . \quad . \quad . \quad \text{from equation (3.1)}$$

$$= \tfrac{1}{2} \times 10^3 \times 9{\cdot}81 \times 2 \times 5{\cdot}4^2$$

$$= 286 \times 10^3 \text{ N}$$

Similarly, $$P_2 = \tfrac{1}{2} \times 10^3 \times 9{\cdot}81 \times 2 \times 1{\cdot}5^2$$

$$= 22 \times 10^3 \text{ N}$$

Thus the resultant force $P = 286 - 22 = 264$ kN

Each of these forces acts through the centre of pressure of their respective sides. If the resultant forces act at a height H above the base, then, taking moments about the base

$$264H = 286 \times \frac{5{\cdot}4}{3} - 22 \times \frac{1{\cdot}5}{3}$$

$$\therefore \underline{H = 1{\cdot}91 \text{ m}}$$

The arrangement of the gates is shown in Fig. 3.12. The force P is normal to the gate and the reaction R between the gates is normal to the line of contact. Hence, from symmetry, the reaction at the hinges is also R, acting through the point of intersection of the other two forces and inclined at 20° to the gate.

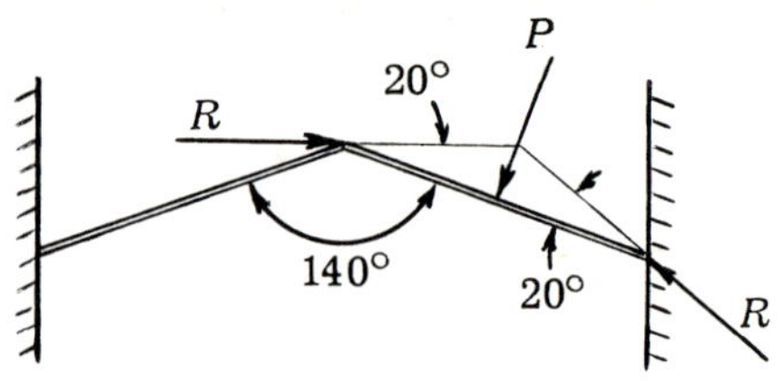

Fig. 3.12

Equating forces normal to the gate,

$$P = 2R \sin 20°$$

from which $R = \underline{386 \text{ kN}}$, acting at the same height above the base as P, i.e. 1·91 m.

If the reactions at the top and bottom hinges are R_1 and R_2 respectively,

then $$R_1 + R_2 = R = 386 \text{ kN} \qquad (1)$$

Taking moments about the base,

$$R_1 \times 5{\cdot}4 + R_2 \times 0{\cdot}6 = R \times 1{\cdot}91 \qquad (2)$$

From equations (1) and (2), $R_1 = \underline{105 \text{ kN}}$ and $R_2 = \underline{281 \text{ kN}}$

3. *The water face of a dam is vertical for 6 m below the water surface and below this level, slopes at 30° to the vertical. The base is 20 m below the surface. Find the magnitude, direction and position of the resultant water thrust per metre width on the face of the dam.*

The arrangement is shown in Fig. 3.13.

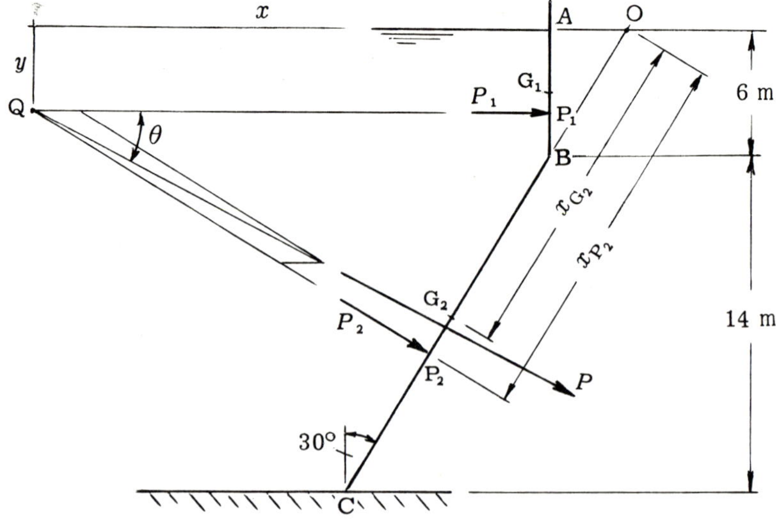

Fig. 3.13

$$P_1 = \tfrac{1}{2}\rho g b h^2 \quad . \quad . \quad . \quad . \quad . \quad \text{from equation (3.1)}$$

$$= \tfrac{1}{2} \times 10^3 \times 9{\cdot}81 \times 1 \times 6^2 \ \text{N}$$

$$= 177 \ \text{kN} \text{, acting at a depth of 4 m below the surface}$$

$$\text{Depth of } G_2 = 6 + 7 = 13 \ \text{m}$$

$$\text{Length BC} = 14 \sec 30^\circ = 16{\cdot}17 \ \text{m}$$

$$x_{G_2} = 13 \sec 30^\circ = 15{\cdot}01 \ \text{m}$$

$$P_2 = \text{area} \times \text{pressure at } G_2$$

$$= 1 \times 16{\cdot}17 \times 10^3 \times 9{\cdot}81 \times 13 \ \text{N}$$

$$= 2\,062 \ \text{kN}$$

$$x_{P_2} = \frac{I_0}{A x_{G_2}}$$

$$= \frac{\dfrac{1 \times 16{\cdot}17^3}{12} + 1 \times 16{\cdot}17 \times 15{\cdot}01^2}{1 \times 16{\cdot}17 \times 15{\cdot}01} = 16{\cdot}46 \ \text{m}$$

From the vector diagram at Q,

$$P = \underline{2\,217 \ \text{kN}}$$

and $$\theta = \underline{27{\cdot}72^\circ}$$

By calculation of scale drawing, the coordinates of the point Q are

$$x = \underline{27{\cdot}53 \ \text{m}}$$

and $$y = \underline{4 \ \text{m}}$$

4. *The sector gate shown in Fig. 3.14 (a) is used to control the depth of water in a canal. The gate profile forms part of a circular arc of radius 2 m and 15 m wide. Determine the thrust on the gate.*

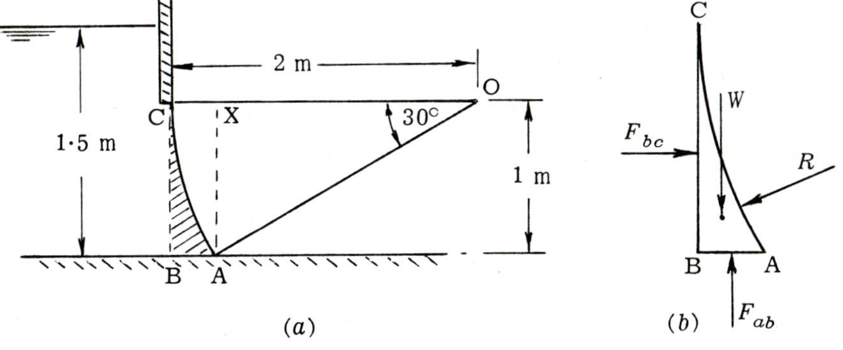

Fig. 3.14

Consider the equilibrium of the wedge of water ABC shown in Fig. 3.14(*b*).

$$\text{Force on BC, } F_{bc} = \text{area} \times \text{pressure at centroid}$$

$$= (1 \times 15) \times (1 \times 10^3 \times 9{\cdot}81) = 147\,000 \text{ N}$$

$$\text{Force on AB, } F_{ab} = \text{area} \times \text{pressure at centroid}$$

$$= (2 - 2\cos 30^\circ) \times 15 \times (1{\cdot}5 \times 10^3 \times 9{\cdot}81) = 59\,300 \text{ N}$$

Weight of liquid wedge ABC,

$$W = \text{weight of OXA} + \text{weight of XCBA} - \text{weight of OCA}$$

$$= 15 \times 10^3 \times 9{\cdot}81 \left\{ \tfrac{1}{2} \times 2\cos 30^\circ \times 1 + (2 - 2\cos 30^\circ) \times 1 - \frac{\pi}{12} \times 2^2 \right\}$$

$$= 12\,800 \text{ N}$$

Therefore resultant vertical force,

$$F_{ab} - W = 59\,300 - 12\,800 = 46\,500 \text{ N}$$

The resultant force on the gate is equal and opposite to the reaction of the gate, R, given by

$$R = \sqrt{F_{bc}^2 + (F_{ab} - W)^2}$$

$$= \sqrt{147\,000^2 + 46\,500^2} = \underline{154\,200 \text{ N}}$$

R acts at an angle $\tan^{-1} \dfrac{46\,500}{147\,000} = 17{\cdot}6^\circ$ to the horizontal. The line of action of R must pass through the hinge O since it is normal to the circular surface. It therefore exerts no torque about O and the gate opening is unaffected by the depth of water.

5. *A stream is spanned by a bridge which is a single mass-concrete arch in the form of a circular arc of 5 m radius, the crown being 2·5 m above the springings. Measured in the direction of the stream, the overall width of the bridge is 6 m.*

During a flood, the water level rises to the level of the crown. Assuming that the arch remains watertight, calculate the force tending to lift the bridge from its foundations.

The upthrust on the bridge, P, is the same as the force which would be required to support a mass of water occupying the area shown shaded in Fig. 3.15 if the bridge were removed.

$$\text{Width of bridge} = 2\sqrt{5^2 - 2{\cdot}5^2} = 8{\cdot}66 \text{ m}$$

$$\cos\theta = \frac{2{\cdot}5}{5} \qquad \therefore \ \theta = 60^\circ$$

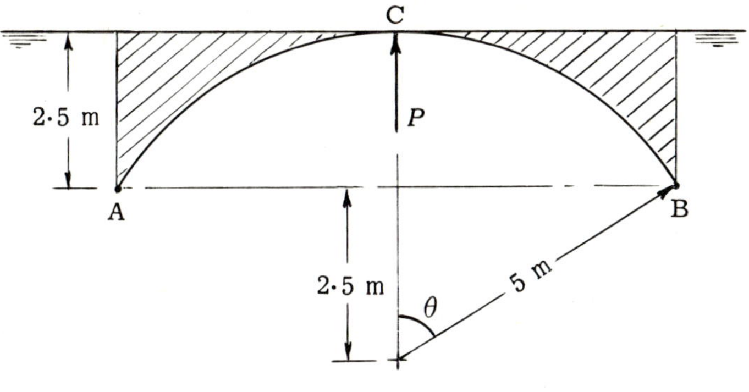

Fig. 3.15

$$\therefore \text{ area of segment ABC} = \frac{120}{360} \times \pi \times 5^2 - \tfrac{1}{2} \times 8{\cdot}66 \times 2{\cdot}5$$

$$= 15{\cdot}355 \text{ m}^2$$

$$\therefore \text{ shaded area} = 8{\cdot}66 \times 2{\cdot}5 - 15{\cdot}355$$

$$= 6{\cdot}295 \text{ m}^2$$

$\therefore$ weight of water displaced by arch

$$= 6{\cdot}295 \times 6 \times 10^3 \times 9{\cdot}81$$

i.e. $P = 370\,500$ N or $\underline{370{\cdot}5 \text{ kN}}$

6. *A diesel engine in a lorry has a power output of 200 kW at sea level, where the air density is 1·225 0 kg/m³, the pressure is 101·3 kN/m² and the temperature is 15°C. What will be the power output of the engine when crossing a mountain pass 5 000 m above sea level, assuming that the power output depends on the mass of air induced.*

From equation (3.5), $p = 101{\cdot}3(1 - 0{\cdot}000\,022\,6h)^{5{\cdot}255}$

Hence, when $h = 5\,000$ m, $p = 53{\cdot}95$ kN/m²

At 5 000 m with a lapse rate of 0·006 5°C/m,

$$T = (273 + 15) - 5\,000 \times 0{\cdot}006\,5 = 255{\cdot}5 \text{ K}$$

From the ideal gas equation

$$pV = mRT\,^*$$ where R is the specific gas constant

Hence $$m = \frac{pV}{RT}$$

* See *Thermodynamics 3/4*, Bacon and Stephens, Art. 2.4.

At sea level, $m = \dfrac{101{\cdot}3V}{288R} = 0{\cdot}352\dfrac{V}{R}$

At 5 000 m, $m = \dfrac{53{\cdot}95V}{255{\cdot}5R} = 0{\cdot}211\dfrac{V}{R}$

Therefore power at 5 000 m $= \dfrac{0{\cdot}211}{0{\cdot}352} \times 200 = \underline{120 \text{ kW}}$

7. A closed cylindrical tank 0·6 m diameter and 1·8 m deep, with vertical axis, contains water to a depth of 1·2 m. Air to a pressure of 40 kN/m² above atmospheric is pumped into the cylinder. Determine the total normal force on the vertical wall of the tank and the distance of the centre of pressure from the base.
(*Ans.*: 149 kN; 0·855 m)

8. A circular opening, 0·6 m diameter, in the vertical side of a tank is closed by a disc which just fits the opening and can rotate about an axis lying along a horizontal diameter.

Calculate: (*a*) the force on the disc, and (*b*) the torque required to maintain the disc in equilibrium in the vertical position when the head of water above the axis of rotation is 0·8 m.

Show that this torque is independent of the head, provided that the disc is always submerged. (*Ans.*: 2·22 kN; 62·4 N m)

9. The outlet of a horizontal rectangular channel, 1·2 m wide by 0·9 m deep is closed by a plane flap inclined at 40° to the vertical and hinged at the horizontal upper edge of the channel. The flap weighs 2·5 kN, the line of action of the weight being 0·35 m horizontally from the hinge.

To what height must the water level in the channel rise so as just to cause the flap to open? (*Ans.*: 0·33 m)

10. A channel of trapezoidal section is 2 m wide at the bottom. It contains water to a depth of 2·5 m and at the surface, the width is 4 m. The channel is closed by a dam, the water face of which is inclined at 30° to the vertical. Calculate the water force on the face of the dam and find the position of the centre of pressure.
(*Ans.*: 13·63 kN; 1·804 m)

11. A rectangular opening in the sloping side of a fresh water reservoir is 3 m deep by 2 m wide, the 2 m side being horizontal. The opening is filled by a door of uniform thickness, hinged at the top and the door is kept closed by its own weight and by a load L on the lever arm shown in Fig. 3.16. The weight of the door is 45 kN. Calculate the load L so that the gate commences to open when the water level is 0·8 m above the top of the gate. (*Ans.*: 49·43 kN)

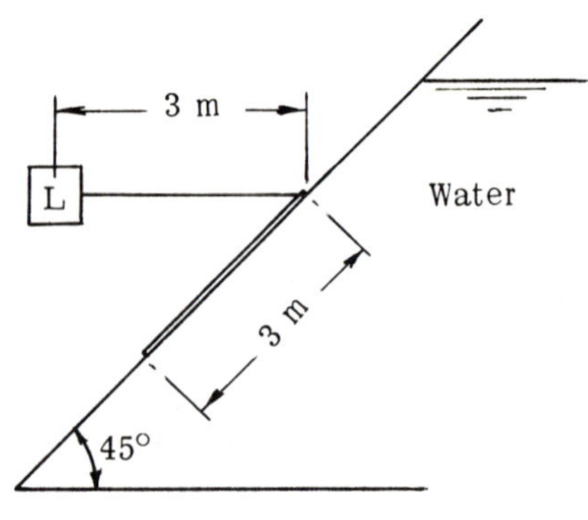

Fig. 3.16

12. The vertical cross-section of a water tank in a barge is shown in Fig. 3.17, the section being symmetrical about the vertical axis. Calculate the magnitude and position of the force on a vertical surface dividing two tanks in the barge when one tank is full and the other is filled to one quarter of its depth.

(*Ans.*: 3·09 MN acting at 4·75 m from the top)

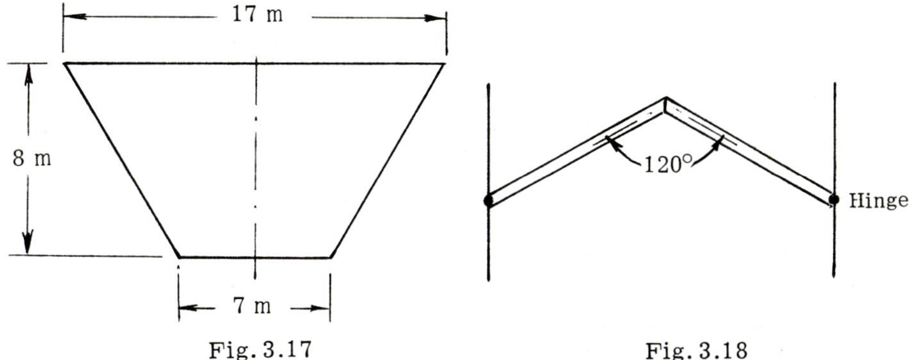

Fig. 3.17 Fig. 3.18

13. Fig. 3.18 shows the gates of a lock which is 8 m wide. The gates make an angle of 120° with each other and each gate is hinged to the side of the lock. Determine the force on each hinge and the thrust between the gates when the depth of water above the gates is 12 m and that below the gates is 4 m.

(*Ans.*: 2·9 MN; 2·9 MN)

14. The sector gate shown in Fig. 3.19 is used to control a water level. The circular arc has a radius of 4 m and the gate is 8 m wide. The gate is pivoted at the centre of the arc, the pivot being 2 m above the bottom of the gate. Determine the thrust on the pivot when the water depth is 2 m.

(*Ans.*: 167 kN at 20·35° to horizontal)

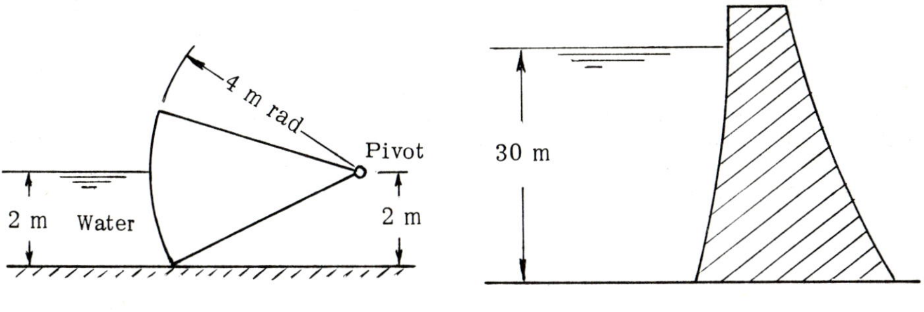

Fig. 3.19 Fig. 3.20

15. Fig. 3.20 shows the vertical cross-section of a dam. The water face is part of a circular arc of radius 120 m, the centre of curvature being 30 m above the base of the dam. Calculate the magnitude and direction of the force on the 100 m wide face when the depth of water is 30 m.

(*Ans.*: 449 MN at 10·1° to the horizontal, through the centre of curvature)

16. Fig. 3.21 shows the vertical cross-section of the side of a ship. Determine the force on the section AB which is 12 m long and has a circular profile of radius 10 m. The density of sea water is 1 026 kg/m³.

(*Ans.*: 14·6 MN at 54·6° to horizontal, through centre of curvature)

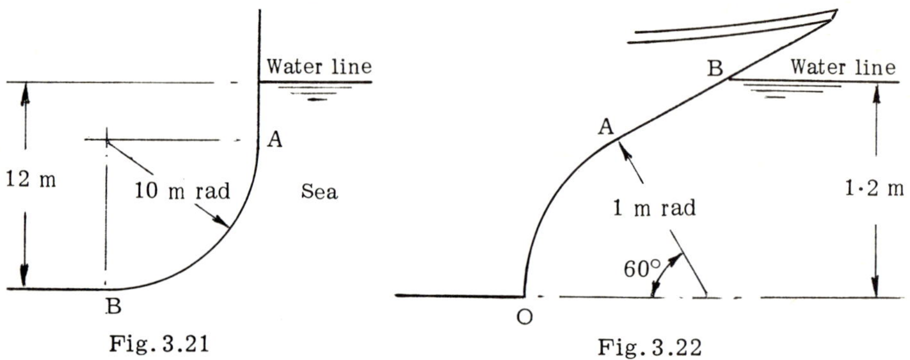

17. The stern of a yacht is shown in Fig. 3.22, OA is a circular arc, AB is a straight line and the stern is 1·3 m wide. Determine the force on the stern for the water level shown, taking the density of sea water as 1 026 kg/m³. (A partly graphical solution is recommended.)

(*Ans.*: 10·6 kN at 28° to the horizontal, intersecting OAB at 0·675 m below the water line)

18. A hinged hatch in a submarine is to be opened by compressed air. The hatch is circular with a diameter of 0·8 m and is inclined at 45° to the vertical. The centre of the hatch is 7 m below the sea surface and the hinge is at the uppermost point. The mass of the hatch is 95 kg and the barometer reads 101 kN/m². Calculate the air pressure required to open the hatch, taking the density of sea water as 1 026 kg/m³. (*Ans.*: 175 kN/m²)

19. The pressure and temperature of air at sea level are 101 kN/m² and 288 K respectively. Determine the pressure, density and temperature at an altitude of 9 000 m, assuming a polytopic atmosphere for which $p/\rho^{1\cdot235}$ = constant.

What lapse rate does this represent?

(*Ans.*: 30·7 kN/m²; 0·466 kg/m³; 229·5 K; 0·006 5 K/m)

20. A motorist decides to convert an aneroid barometer to an altimeter to use in his car on a mountain holiday. The barometer scale is marked at 900, 950 and 1 000 mb. What heights will correspond to the 900 and 950 mb readings assuming that at sea level, where the air density is 1·225 kg/m³, the average barometer reading is 1 000 mb? (*Ans.*: 898 m; 432 m)

21. An airfield is situated on a plateau 3 000 m above sea level. For landings at this airfield, the altimeter has to be offset to allow for this elevation, the altimeter setting being determined by the pressure at local ground level. Assuming sea level conditions to be 101·32 kN/m², 15·2°C, 1·225 kg/m³, determine the pressure at the airfield. (*Ans.*: 70·1 kN/m²)

4 Flow and flow measurement

4.1 Continuity equation When a liquid is flowing along a pipe, the motion is normally turbulent and individual particles do not follow a regular path. The axial velocity is approximately uniform across the section except very close to the pipe wall and it is usual to assume a uniform velocity across the whole section.

Since liquids are assumed to be incompressible, the volume flow rate Q, in the pipe must be constant. Thus, in the tapered pipe shown in Fig. 4.1,

$$Q = A_1 V_1 = A_2 V_2 \qquad (4.1)$$

Equation (30.1) is known as the *continuity equation.*

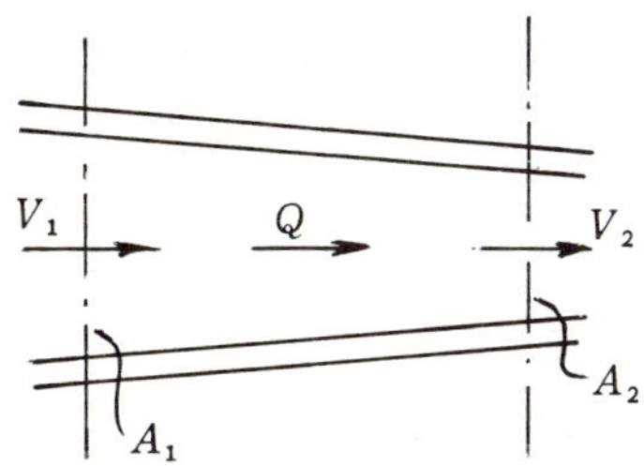

Fig. 4.1

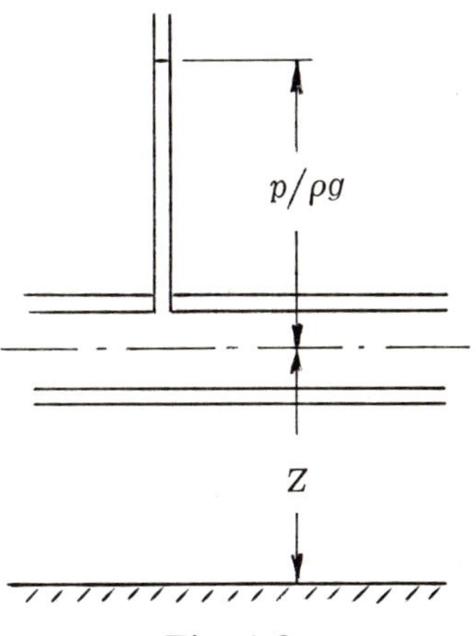

Fig. 4.2

4.2 Energy of a moving liquid Consider a liquid at a pressure p, moving with a velocity V at a height Z above datum level, Fig. 4.2.

If a tube were inserted in the top of the pipe, the liquid would rise up the tube a distance $p/\rho g$ and this is equivalent to an additional height of liquid relative to datum level.

$$\text{Thus potential energy of liquid per unit mass} = g\left(Z + \frac{p}{\rho g}\right)$$

$$\text{and kinetic energy of liquid per unit mass} = \frac{V^2}{2}$$

$$\text{Therefore total energy of liquid per unit mass} = g\left(Z + \frac{p}{\rho g} + \frac{V^2}{2g}\right) \qquad (4.2)$$

Equation (4.2) represents the specific energy of the liquid. Each of the quantities in the brackets have the units of length and are termed *heads.* Thus Z is referred to as the *potential head* of the liquid, $p/\rho g$ as the *pressure head* and $V^2/2g$ as the *velocity head.* The expression $Z + p/\rho g + V^2/2g$ is therefore the total head of the liquid and represents the energy per unit weight.

If there are no losses of energy between two sections of a pipe, Fig. 4.3, and no energy changes due to work or heat transfer, the total energy remains constant,

i.e.
$$Z_1 + \frac{p_1}{\rho g} + \frac{V_1^2}{2g} = Z_2 + \frac{p_2}{\rho g} + \frac{V_2^2}{2g} \tag{4.3}$$

This is known as *Bernoulli's equation.*

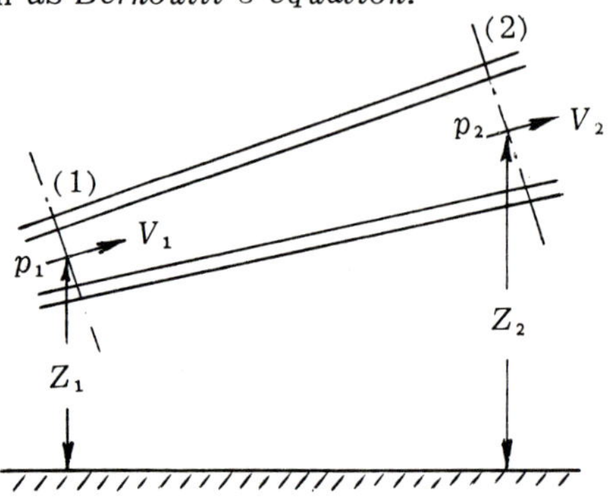

Fig. 4.3

If a loss of head h occurs between sections (1) and (2), equation (4.3) becomes

$$Z_1 + \frac{p_1}{\rho g} + \frac{V_1^2}{2g} = Z_2 + \frac{p_2}{\rho g} + \frac{V_2^2}{2g} + h \tag{4.4}$$

If, in addition, energy is added by a pump or deducted by a turbine, equation (4.3) becomes further modified to

$$Z_1 + \frac{p_1}{\rho g} + \frac{V_1^2}{2g} + \frac{w_{\text{in}}}{g} = Z_2 + \frac{p_2}{\rho g} + \frac{V_2^2}{2g} + \frac{w_{\text{out}}}{g} + h \tag{4.5}$$

where w is the specific work in J/kg.

4.3 Venturi meter Fig. 4.4 shows a venturi meter, which consists of a constriction in a pipe with a manometer arranged to indicate the difference in pressure between the main and the throat.

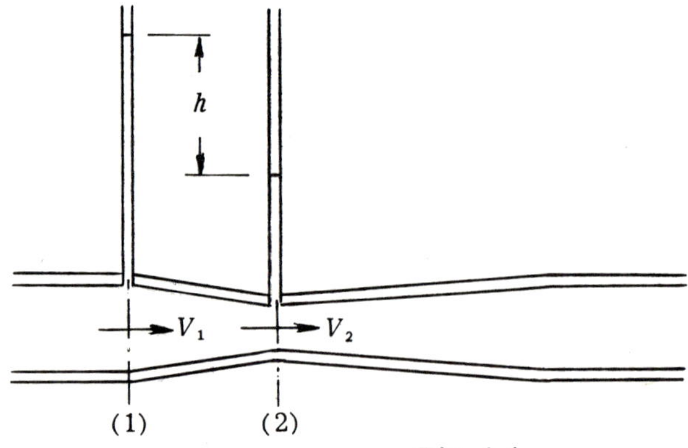

Fig. 4.4

Applying Bernoulli's equation to sections (1) and (2) and assuming no losses,

$$Z_1 + \frac{p_1}{\rho g} + \frac{V_1^2}{2g} = Z_2 + \frac{p_2}{\rho g} + \frac{V_2^2}{2g}$$

i.e. $$\frac{p_1}{\rho g} - \frac{p_2}{\rho g} = \frac{V_2^2}{2g} - \frac{V_1^2}{2g}$$ if the meter is horizontal.

But $$A_1 V_1 = A_2 V_2$$

$$\therefore \quad V_2 = \frac{A_1}{A_2} V_1$$

$$\therefore \quad \frac{p_1}{\rho g} - \frac{p_2}{\rho g} = h = \frac{V_1^2}{2g}\left(\frac{A_1}{A_2}\right)^2 - \frac{V_1^2}{2g} = \frac{V_1^2}{2g}\left(\frac{A_1^2 - A_2^2}{A_2^2}\right)$$

$$\therefore \quad V_1 = \frac{A_2}{\sqrt{A_1^2 - A_2^2}}\sqrt{2gh}$$

$$\therefore \quad Q = A_1 V_1 = \frac{A_1 A_2}{\sqrt{A_1^2 - A_2^2}}\sqrt{2gh}$$

$$= k\sqrt{h} \qquad \text{since } A_1 \text{ and } A_2 \text{ are constants} \qquad (4.6)$$

Due to friction in the tapered pipe, the pressure at the throat is slightly lower than the theoretical value, so that h becomes slightly larger, giving a value for Q which is too high. To correct this, the theoretical discharge is multiplied by a coefficient of discharge, c_d, which is usually about 0·97 or 0·98.

Thus $$Q = c_d k\sqrt{h} \qquad (4.7)$$

The difference in pressure between the main and the throat is usually measured with a mercury-and-water U-tube, Art. 2.3(*c*), so that

$$h = x(S - 1)$$

If the venturi meter is inclined to the horizontal such that the sections (1) and (2) are at heights Z_1 and Z_2 above datum level, equation (4.6) becomes

$$Q = k\sqrt{h - (Z_2 - Z_1)}$$

If the pressure difference is measured with a mercury-and-water U-tube, Fig. 4.5, then, equating pressures at level XX,

$$\frac{p_1}{\rho g} + Z_1 = \frac{p_2}{\rho g} + (Z_2 - x) + Sx$$

i.e. $$h - (Z_2 - Z_1) = x(S - 1)$$

Thus, for a given flow rate, the manometer reading is the same for an inclined meter as for a horizontal meter.

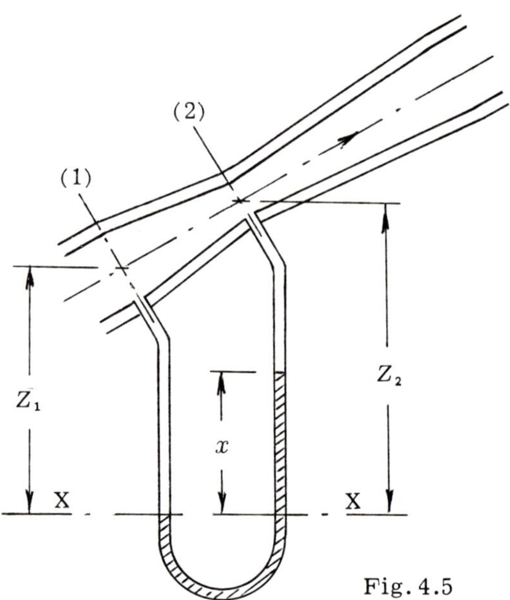

Fig. 4.5

4.4 Small orifices An orifice is a precisely made hole through which a liquid may flow. A *small* orifice is one in which the variation of head across the hole is small enough to be neglected.

Fig. 4.6(a) shows a small sharp-edged orifice through which a liquid issues under a head h. The theoretical velocity of the jet is obtained by applying Bernoulli's equation to points (1) and (2),

i.e.
$$h + \frac{p_1}{\rho g} + 0 = 0 + \frac{p_2}{\rho g} + \frac{V^2}{2g}$$

But
$$p_1 = p_2 = \text{atmospheric pressure}$$

Hence
$$V = \sqrt{2gh}$$

Due to the component of the velocity perpendicular to the axis, the jet will contract after leaving the orifice and the section at which the jet becomes parallel is termed the *vena contracta*.

The ratio $\dfrac{\text{area of vena contracta}}{\text{area of orifice}}$ is called the *coefficient of contraction* (c_c).

Due to friction at the orifice, the velocity at the vena contracta is slightly less than the theoretical velocity and the ratio $\dfrac{\text{actual velocity of jet}}{\text{theoretical velocity of jet}}$ is called the *coefficient of velocity* (c_v).

If a is the orifice area, the discharge, or flow rate,

$$Q = c_c a \times c_v \sqrt{2gh}$$

$$= c_d a \sqrt{2gh} \quad \text{where} \quad c_d = c_c c_v \qquad (4.8)$$

c_d is the *coefficient of discharge* which is the ratio $\dfrac{\text{actual discharge}}{\text{theoretical discharge}}$

For a sharp-edged orifice, approximate values for c_c, c_v and c_d are 0·64, 0·97 and 0·62.

The value for c_d can be improved by fitting a pipe or mouthpiece to the orifice, as shown in Figs. 4.6(b) and (c). .

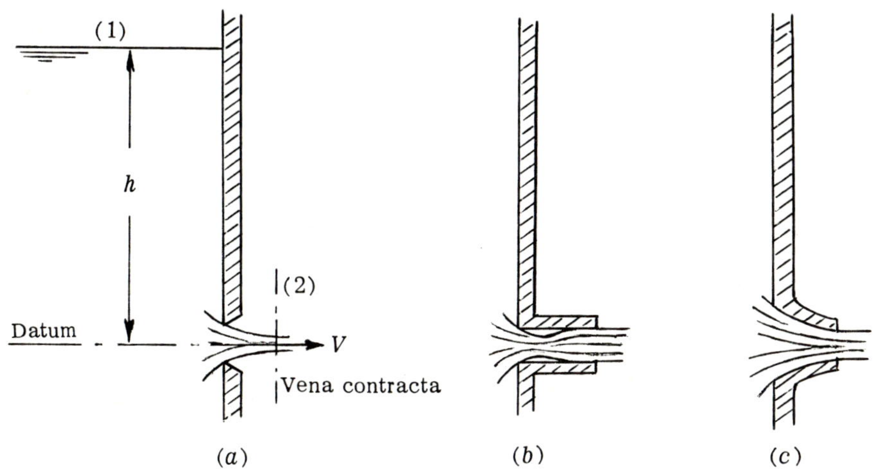

Fig. 4.6

4.5 Large rectangular orifice For the orifice shown in Fig. 4.7, there will be an appreciable difference in pressure between the top and bottom edges. If the breadth of the orifice is b, the theoretical discharge through an elementary strip of thickness dh at a depth h below the surface is given by

$$dQ = b\,dh \times \sqrt{2gh}$$

$$\therefore \quad Q = b\sqrt{2g}\int_{h_1}^{h_2} h^{1/2}\,dh$$

$$= \tfrac{2}{3}\, b\sqrt{2g}\,(h_2^{3/2} - h_1^{3/2})$$

$$\text{Therefore actual discharge} = c_d \times \tfrac{2}{3}\, b\sqrt{2g}\,(h_2^{3/2} - h_1^{3/2}) \qquad (4.9)$$

4.6 Orifice plate Fig. 4.8 shows a sharp-edged orifice plate fitted into a pipe conveying a liquid. The flow will follow the pattern shown, forming a vena contracta just beyond the orifice before expanding to the pipe diameter again.

Applying Bernoulli's equation to sections (1) and (2),

$$\frac{p_1}{\rho g} + \frac{V_1^2}{2g} = \frac{p_2}{\rho g} + \frac{V_2^2}{2g}$$

Also $\qquad A_1 V_1 = A_2 V_2 \qquad$ so that $\quad V_2 = \dfrac{A_1}{A_2} V_1$

Thus $$\frac{p_1}{\rho g} - \frac{p_2}{\rho g} = h = \frac{V_1^2}{2g}\left(\frac{A_1^2 - A_2^2}{A_2^2}\right)$$

Therefore $$Q = A_1 V_1 = \frac{A_1 A_2}{\sqrt{A_1^2 - A_2^2}}\sqrt{2gh} = k\sqrt{h} \qquad (4.10)$$

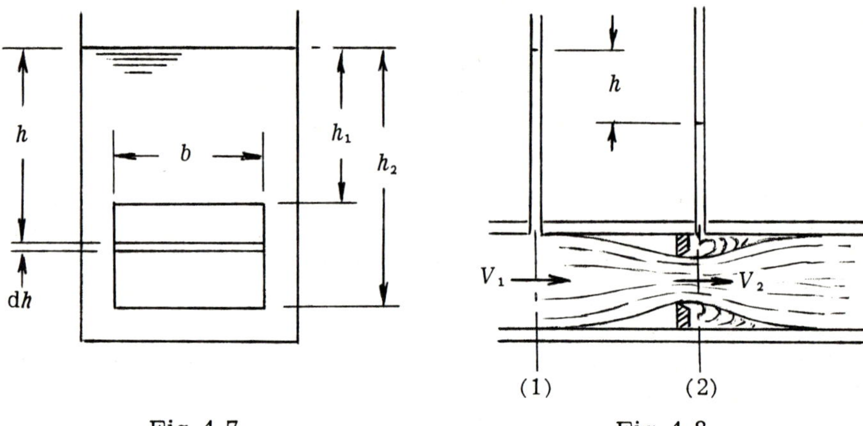

Fig. 4.7

Fig. 4.8

It is difficult to establish the exact position of the vena contracta and to determine its area, A_2, in relation to the known orifice area. The constant k is therefore determined experimentally, when it will incorporate the coefficient of discharge.

If h is small so that ρ is approximately constant, equation (4.10) may also be applied to compressible fluid flow. This type of orifice is used to meter flow in carburettors, fuel systems, control systems and engine test installations.

4.7 Pitot tube A simple pitot tube consists of a tube, bent at right angles and tapered at the lower end, as shown in Fig. 4.9. If such a tube is placed in an open channel, the liquid at the mouth of the tube is brought to rest (this point being termed a *stagnation point*) and liquid is forced up the tube above the level of the free surface.

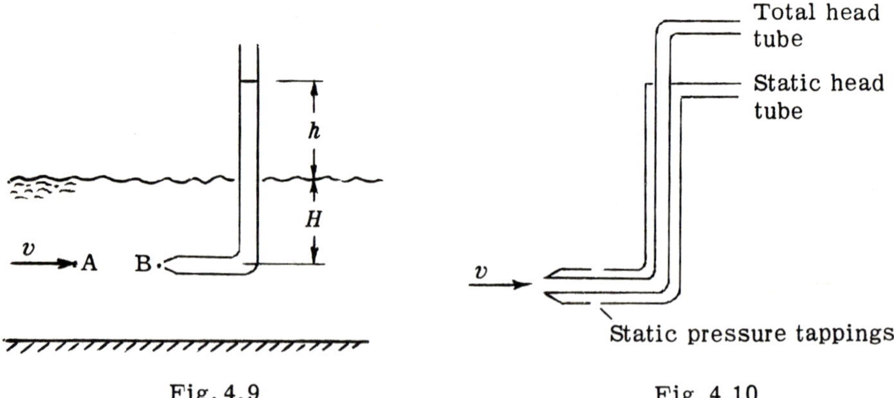

Fig. 4.9

Fig. 4.10

Applying Bernoulli's Equation to a point A in the free stream and a point B at the mouth of the pitot tube,

$$H + \frac{v^2}{2g} = H + h$$

or $$v = \sqrt{2gh} \tag{4.11}$$

If the pitot tube is used in a pipe under pressure, the reading h will represent both the velocity head and pressure head at the point under consideration and it is necessary to deduct the pressure head in order to obtain the velocity head only.

This is conveniently arranged by using a *pitot-static tube*, Fig. 4.10. This consists of two concentric tubes, the inner one only being open at the lower end. The outer tube has radial holes drilled in it to measure the static pressure head, while the inner tube measures the total head. The two outlets are connected to a U-tube manometer of the type shown in Fig. 2.7 and the difference in levels represents the velocity head only, from which the velocity can be deduced.

If the pitot tube is large enough to distort the flow, a calibration constant C may be required, so that

$$v = C\sqrt{2gh} \tag{4.12}$$

Pitot tubes may also be used to measure gas velocity if the velocity is sufficiently low so that the density may be considered constant. They can also be used to determine the velocities of aircraft and ships.

4.8 Velocity distribution and flow rate A pitot tube measures the velocity at a particular point in the section of a channel or pipe and since the velocity varies considerably over the cross-section, a series of pitot readings is necessary to determine the flow rate.

Fig. 4.11 shows the type of variation which may be obtained by traversing the pitot tube along the line AA. The cross-section is then divided into convenient areas a_1, a_2, a_3, etc., Fig. 4.12, and the velocity at the centre of each area is determined.

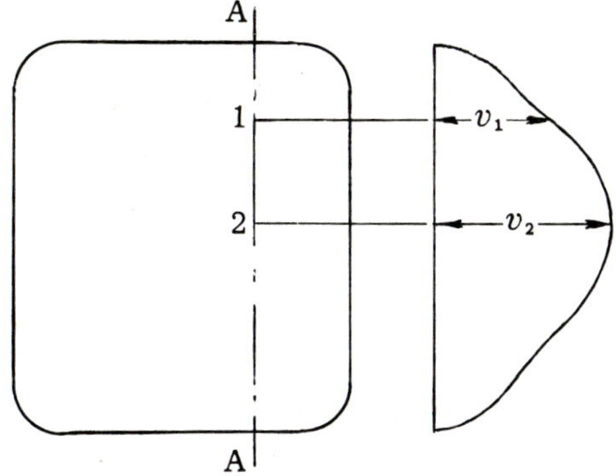

Fig. 4.11

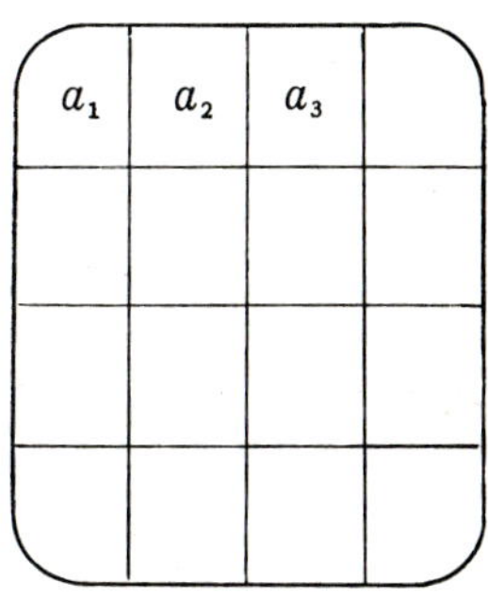

Fig. 4.12

Then the flow rate

$$Q = a_1v_1 + a_2v_2 + a_3v_3 + \ldots = \Sigma av \tag{4.13}$$

If the total cross-sectional area of the duct is A, the mean velocity

$$\text{the mean velocity } V = \frac{Q}{A} = \frac{\Sigma av}{A} \tag{4.14}$$

In the case of a circular pipe, the velocity profile is the same across any diameter and hence, for an annular ring of mean radius r and area Δa, Fig. 4.13,

$$\Delta Q = v_r \times \Delta a \quad ,$$

where v_r is the velocity at radius r

Thus
$$Q = \Sigma v_r \Delta a \tag{4.15}$$

and
$$V = \frac{\Sigma v_r \Delta a}{\pi r^2} \tag{4.16}$$

The velocity at the boundary of any duct or pipe is zero.

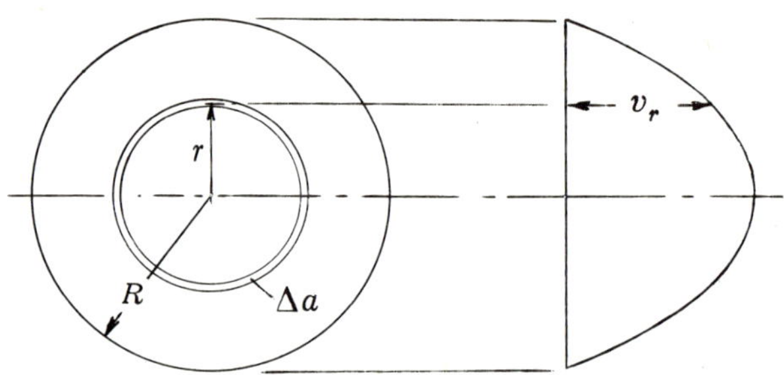

Fig. 4.13

1. *Fig. 4.14 shows part of the lubrication system of an engine. At A the pressure is atmospheric and at B the gauge pressure is 80 kN/m². The diameter of the suction pipe is twice that of the delivery pipe. The total head lost in the system is 440 mm of oil and point B is 0·8 m above A. The pump supplies a head of 11·3 m of oil and the oil has a density of 850 kg/m³.*

Determine the velocity of flow in the suction pipe.

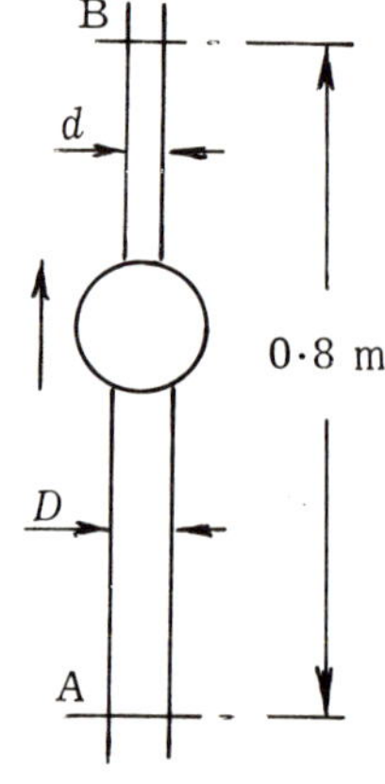

Fig. 4.14

From equation (4.8),

$$Z_a + \frac{p_a}{\rho g} + \frac{V_a^2}{2g} + \frac{w_{\text{in}}}{g} = Z_b + \frac{p_b}{\rho g} + \frac{w_{\text{out}}}{g} + h$$

Also $\quad Q = A_a V_a = A_b V_b$

$$\therefore \; V_b = V_a\left(\frac{D^2}{d^2}\right) = 4V_a$$

Hence
$$0 + 0 + \frac{V_a^2}{2g} + 11{\cdot}3 = 0{\cdot}8 + \frac{80 \times 10^3}{850g} + \frac{(4V_a)^2}{2g} + 0 + \frac{440}{10^3}$$

$$\therefore \; \frac{V_a^2}{2 \times 9{\cdot}81}(16 - 1) = 0{\cdot}46$$

from which
$$V_a = \underline{0{\cdot}775 \text{ m/s}}$$

2. *A horizontal venturi meter with a throat diameter of 50 mm is placed in a pipe of 120 mm diameter. The pipe conveys oil of density 800 kg/m³ and the difference in levels of mercury in a U-tube manometer is 220 mm. The coefficient of discharge of the meter is 0·98. Calculate the flow rate of oil.*

From Art. 4.3,
$$Q = c_d \frac{A_1 A_2}{\sqrt{A_1^2 - A_2^2}} \sqrt{2gh}$$

where h is the difference in head of liquid between the main and the throat.

From equation (2.7),
$$h = \frac{p_1 - p_2}{\rho g} = \left(\frac{\rho_{Hg}}{\rho_{oil}} - 1\right)x$$
$$= \left(\frac{13{\cdot}6}{0{\cdot}8} - 1\right) \times 0{\cdot}22 = 3{\cdot}52 \text{ m}$$

Hence
$$Q = 0{\cdot}98 \frac{\frac{\pi}{4} \times 0{\cdot}12^2 \times \frac{\pi}{4} \times 0{\cdot}05^2}{\sqrt{\left(\frac{\pi}{4} \times 0{\cdot}12^2\right)^2 - \left(\frac{\pi}{4} \times 0{\cdot}05^2\right)^2}} \sqrt{2 \times 9{\cdot}81 \times 3{\cdot}52}$$
$$= \underline{16{\cdot}2 \times 10^{-3} \text{ m}^3/\text{s}}$$

3. *In an engine test, 0·04 kg/s of air flows through a 50 mm diameter pipe, into which is fitted a 40 mm diameter orifice plate. The density of air is 1·2 kg/m³ and the coefficient of discharge for the orifice is 0·63. The pressure drop across the orifice is measured by a U-tube manometer filled with water.*

Calculate the manometer reading.

The arrangement is shown in Fig. 4.15.

Mass flow rate, $\dot{m} = \rho Q$

i.e. $0{\cdot}4 = 1{\cdot}2Q$

$\therefore\ Q = 0{\cdot}033\,3 \text{ m}^3/\text{s}$

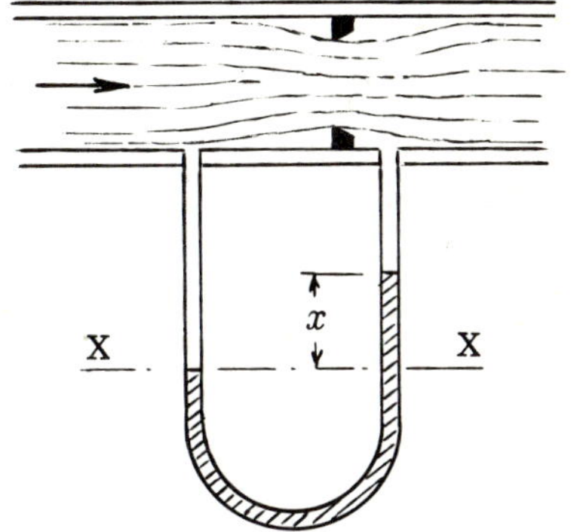

Fig. 4.15

From equation (4.10), the theoretical volume flow rate is given by

$$Q = \frac{A_1 A_2}{\sqrt{A_1^2 - A_2^2}} \sqrt{2gh}$$
$$= \frac{\frac{\pi}{4} \times 0{\cdot}05^2 \times \frac{\pi}{4} \times 0{\cdot}04^2}{\sqrt{\left(\frac{\pi}{4} \times 0{\cdot}05^2\right)^2 - \left(\frac{\pi}{4} \times 0{\cdot}04^2\right)^2}} \sqrt{2 \times 9{\cdot}81 \times h}$$
$$= 0{\cdot}007\,244\sqrt{h}$$

Hence, actual discharge $= 0{\cdot}63 \times 0{\cdot}007\,244\sqrt{h} = 0{\cdot}033\,3$

from which $h = 53{\cdot}34$ m of air

Equating pressures at level XX in the U-tube,

$$1{\cdot}2g \times 53{\cdot}34 = 10^3 g \times x$$

$$\therefore\ x = 64 \times 10^{-3}\ \text{m} = \underline{64\ \text{mm}}$$

4. *A tank is divided into two compartments by a vertical wall in which there is a rectangular orifice 0·8 m wide by 0·5 m deep to enable the oil in the tank to flow from one side to the other. Determine the discharge through the orifice when the level on one side is 3 m above the top of the orifice and that on the other side is 0·2 m below the top of the orifice. The coefficient of discharge is 0·65.*

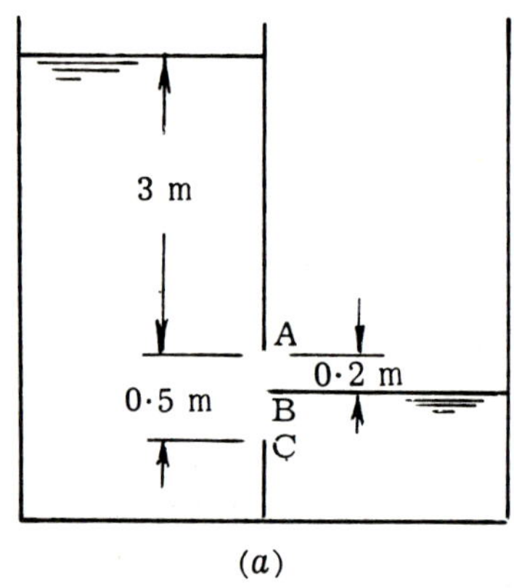

(a)

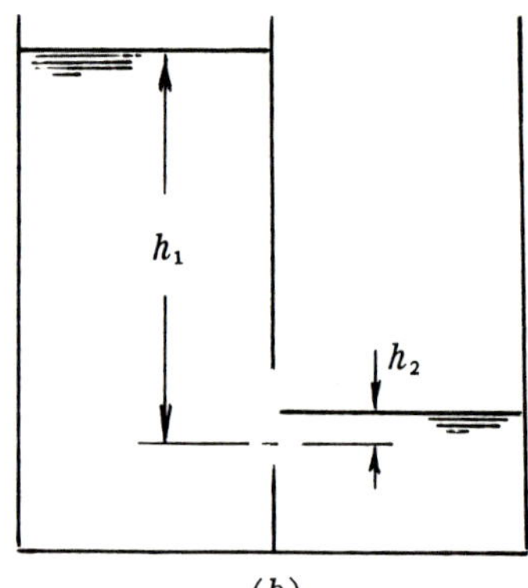

(b)

Fig. 4.16

The upper part of the orifice AB, Fig. 4.16(*a*) permits free flow and the discharge is given by equation (4.9).

Hence
$$Q_{ab} = c_d \times \tfrac{2}{3} b\sqrt{2g}\,(h_2^{3/2} - h_1^{3/2})$$
$$= 0{\cdot}65 \times \tfrac{2}{3} \times 0{\cdot}8\sqrt{2 \times 9{\cdot}81}\,(3{\cdot}2^{3/2} - 3{\cdot}0^{3/2})$$
$$= 0{\cdot}81\ \text{m}^3/\text{s}$$

The lower part, BC, is a 'drowned' orifice and the head causing flow at any point within BC is the difference in level, $h_1 - h_2$, Fig. 4.16(*b*), which is a constant, equal to 3·2 m.

Hence
$$Q_{bc} = c_d A\sqrt{2gH}$$
$$= 0{\cdot}65 \times 0{\cdot}8 \times 0{\cdot}3\sqrt{2 \times 9{\cdot}81 \times 3{\cdot}2}$$
$$= 1{\cdot}25\ \text{m}^3/\text{s}$$

Therefore total discharge $= 0{\cdot}81 + 1{\cdot}25 = \underline{2{\cdot}06\ \text{m}^3/\text{s}}$

5. *(a) A pitot-static tube on the centre line of a 0·36 m diameter circular water pipe records a reading of 0·058 4 m of water. If the meter calibration coefficient is 1, what is the centre line velocity?*

(b) The variation of velocity in the pipe is determined by a pitot-static traverse and the results are shown in the following table. Determine the quantity flowing and the mean velocity in the pipe.

Radius (m)	*0*	*0·06*	*0·12*	*0·15*	*0·17*	*0·18*
Velocity (m/s)	*1·07*	*0·96*	*0·85*	*0·73*	*0·55*	*0*

(*a*) From equation (4.12),

$$v = C\sqrt{2gh} = 1\sqrt{2\times 9{\cdot}81\times 0{\cdot}058\,4} = \underline{1{\cdot}07\ \text{m/s}}$$

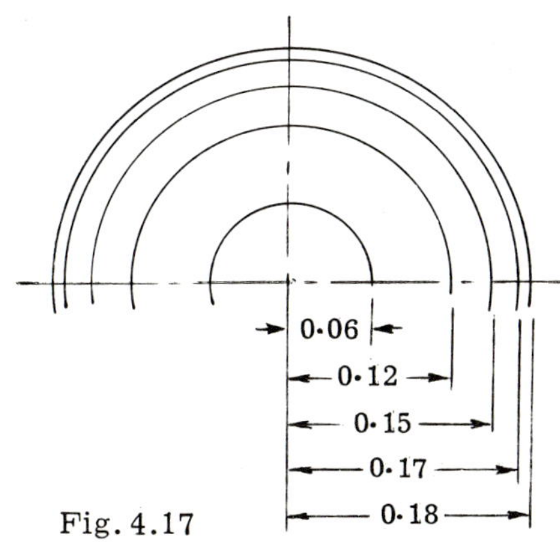

Fig. 4.17

(*b*) The cross-section of the pipe is divided into five concentric rings of radii corresponding with the given information, Fig. 4.17. The mean velocity for any ring is taken as the average of those at the boundaries of the ring. The following table is then compiled:

Mean velocity for ring, v_r	$\frac{1{\cdot}07+0{\cdot}96}{2}$ = 1·015	$\frac{0{\cdot}96+0{\cdot}85}{2}$ = 0·905	$\frac{0{\cdot}85+0{\cdot}73}{2}$ = 0·790	$\frac{0{\cdot}73+0{\cdot}55}{2}$ = 0·640	$\frac{0{\cdot}55+0}{2}$ = 0·275
Area of ring, Δa	0·011 3	0·033 9	0·025 4	0·020 2	0·011 0
$\Delta Q = v_r \Delta a$	0·011 5	0·030 7	0·020 1	0·012 9	0·003 0

$$Q = \Sigma\Delta Q = 0{\cdot}078\,2\ \text{m}^3/\text{s}$$

$$V = \frac{Q}{A} = \frac{0{\cdot}078\,2}{\pi\times 0{\cdot}18^2}$$

$$= \underline{0{\cdot}768\ \text{m/s}}$$

Alternative method The given information is plotted and the cross-section is then divided into smaller rings of uniform width. The mean velocity for each ring is then obtained from the graph, Fig. 4.18.

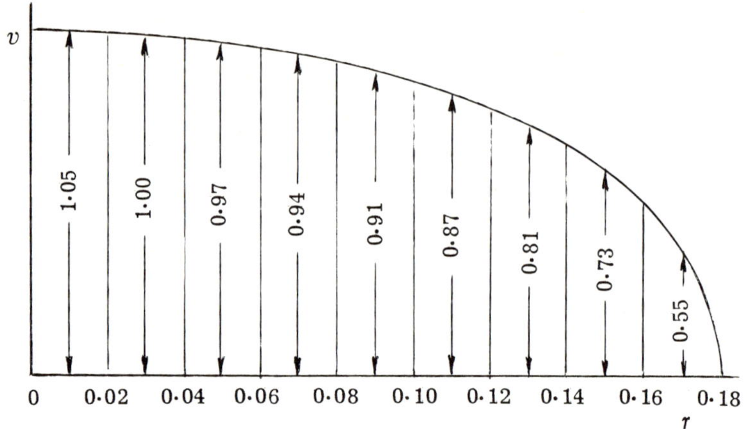

Fig. 4.18

Since the width of each ring is small, the area of a ring may be taken as $2\pi r\,\Delta r$, where r is the mean radius and Δr is the width.

Then $$\Delta Q = v_r \times 2\pi r\,\Delta r$$

and $$Q = 2\pi\,\Delta r\,\Sigma\,v_r r$$

The following table is then compiled:

r	0·01	0·03	0·05	0·07	0·09	0·11	0·13	0·15	0·17
v_r	1·05	1·00	0·97	0·94	0·91	0·87	0·81	0·73	0·55
$v_r r$	0·010 5	0·030 0	0·048 5	0·065 8	0·081 9	0·095 7	0·105 3	0·109 5	0·093 5

$$Q = 2\pi\,\Delta r\,\Sigma\,v_r r$$

$$= 2\pi \times 0{\cdot}02 \times 0{\cdot}641 = \underline{0{\cdot}080\,6\ \text{m}^3/\text{s}}$$

The difference in the answers obtained is due to the approximations inherent in each method. Care is needed to be certain that the approximations made do not give inaccurate answers.

6. A circular pipe has a fall of 4 m and narrows so that, at exit, its area is one quarter that at inlet. The pressure at inlet is 20 kN/m² above atmospheric and at exit, where the velocity is 15 m/s, the pressure is atmospheric. Calculate the inlet velocity and the mass flow rate of water through the pipe per unit cross-sectional area at inlet. (*Ans.*: 2·81 m/s; 2810 kg/s m²)

7. The water tap in the kitchen of a house is 15 mm in diameter and is fed from a tank in the roof. The level in the tank is a constant 6 m above the tap. What is the maximum flow rate through the tap? (*Ans.*: 0·0019 m³/s)

8. A turbine is installed in a vertical pipe line, 150 mm diameter. A gauge is connected to the pipe at a point A above the turbine and another at a point B below the turbine, the vertical distance between A and B being 1·2 m. When the discharge is 0·06 m³/s, the gauges read 400 kN/m² and 35 kN/m² respectively.

Calculate the output power from the turbine, assuming it to be 85% efficient. (*Ans.*: 19·22 kW)

9. In a vertical pipe line conveying water, pressure gauges are inserted at A and B where the diameters are 150 mm and 75 mm respectively. The point B is 2·4 m below A and when the rate of flow down the pipe is 0·02 m^3/s, the pressure at B is 12 kN/m^2 greater than at A. Assuming the losses in the pipe between A and B can be expressed as $k.\frac{v^2}{2g}$, where v is the velocity at A, find the value of k.

(*Ans*.: 3·02)

10. A cylindrical tank 2 m in diameter has a 50 mm diameter orifice in the base, for which the coefficient of discharge is 0·62.

(*a*) Water enters the tank at 0·009 m^3/s. What is the depth in the tank when the level becomes steady?

(*b*) If the input flow rate is increased to 0·02 m^3/s, at what rate does the water level rise when the depth is 4 m? (*Ans*.: 2·8 m; 3 mm/s)

11. Two vertical sided locks are connected by a door 1 m square. The level in one lock is 10 m above the centre of the door and in the other 4 m above the centre. The coefficient of discharge of the door is 0·6. Calculate the flow rate when the door is opened wide. (*Ans*.: 6·53 m^3/s)

12. A venturi meter is to be used to measure the flow of oil in a vertical pipe 150 mm in diameter. The flow is upward through the meter, which has a throat diameter of 50 mm. A mercury manometer connected between the throat and a point in the pipe below the throat reads 160 mm. The coefficient of discharge of the meter is 0·98 and the densities of oil and mercury are 850 kg/m^3 and 13 600 kg/m^3 respectively. Calculate the mass flow rate of oil. (*Ans*.: 11·2 kg/s)

13. A venturi meter is tested with its axis horizontal and the flow is measured in a collecting tank. The pipe diameter is 75 mm, the throat diameter is 38 mm and the pressure difference is measured with a mercury-and-water differential manometer. When the manometer reads 270 mm, 1 100 kg of water were collected in 120 s. Determine the coefficient of discharge of the meter. (*Ans*.: 0·96)

14. The air supply to a petrol engine passes through the carburretor choke tube 40 mm diameter. When the engine is running, the pressure at the throat of the choke tube is 7 N/m^2 below atmospheric. By treating air as an incompressible fluid (the pressure changes are small) and ignoring other losses, calculate the air/fuel ratio by mass used in the engine when the fuel flow rate is 1·25 kg/h. The density of air is 1·2 kg/m^3 and the coefficient of discharge for the choke is 0·95. The velocity of the air away from the choke tube may be assumed negligible. (*Ans*.: 14 : 1)

15. The air supply to an engine on a test bed passes down a 180 mm diameter pipe fitted with an orifice plate 90 mm diameter. The pressure drop across the orifice is 80 mm of paraffin. The coefficient of discharge of the orifice is 0·62 and the densities of air and paraffin are 1·2 kg/m^3 and 830 kg/m^3 respectively. Calculate the mass flow rate of air to the engine. (*Ans*.: 0·1625 kg/s)

16. A pitot static tube is placed on the centre line of a 0·25 m diameter circular pipe carrying air of density 1·1 kg/m^3. The reading of the instrument, which has a calibration coefficient of 0·98, is 40 mm water. If the mean velocity in the pipe is 0·83 of the maximum velocity, determine the volume flow rate. (*Ans*.: 1·066 m^3/s)

17. In a pitot static traverse of a circular duct of 0·5 m diameter the following results were obtained. Determine the volume flow rate of air and the mean velocity.

radius, m	0	0·05	0·10	0·15	0·20	0·25
velocity, m/s	29	28	26	24	14	0

(*Ans*.: 3·39 m^3/s, 17·3 m/s)

18. A pitot static tube is placed at a certain radius in a pipe to measure the velocity of flow. The pipe is 240 mm diameter and conveys 0·2 m^3/s of water. The tube is connected to a differential manometer which reads a pressure difference of 85 mm of mercury. Calculate the calibration coefficient for the pitot tube if it is to be used at this radius to indicate mean velocity. (*Ans*.: 0·9644)

5 Momentum of fluids

5.1 The momentum equation The momentum equation is obtained from Newton's Second Law. When applied to fluid flow, it states that the sum of the external forces acting on a fluid in a given direction is equal to the rate of change of momentum of the fluid in that direction.

Referring to Fig. 5.1, the force in the x direction exerted *by* the surface *on* the fluid is given by

$$F_x = \dot{m}(V_2 \cos \theta_2 - V_1 \cos \theta_1)$$

The force exerted *by* the jet *on* the surface is equal and opposite to the above force.

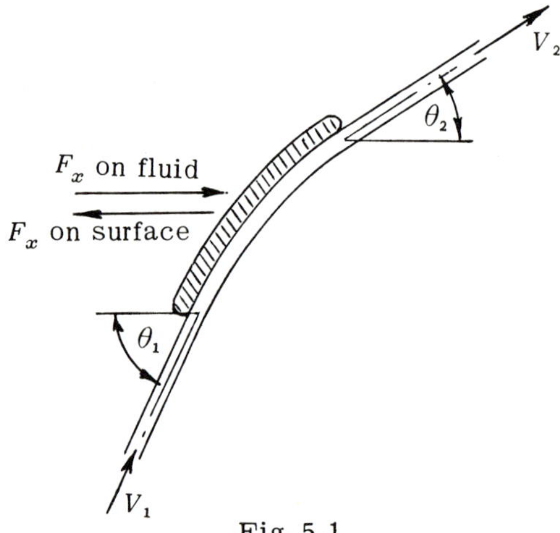

Fig. 5.1

5.2 Impact of jets on stationary surfaces When a jet of liquid impinges on a plate or vane, the force acting on the fluid to change its momentum is the reaction of the vane and the force exerted *by* the jet *on* the vane is equal and opposite to this reaction. In the cases considered below, it will be assumed that there is no friction between the liquid and the vane.

(*a*) *Flat plate normal to jet* Fig. 5.2 shows a jet of cross-sectional area a moving with a velocity V. After impact, the velocity normal to the plate is zero so that force on plate normal to jet

$$= \dot{m}(V - 0)$$

But $\dot{m} = \rho a V$

so that $$F = \rho a V^2 \qquad (5.1)$$

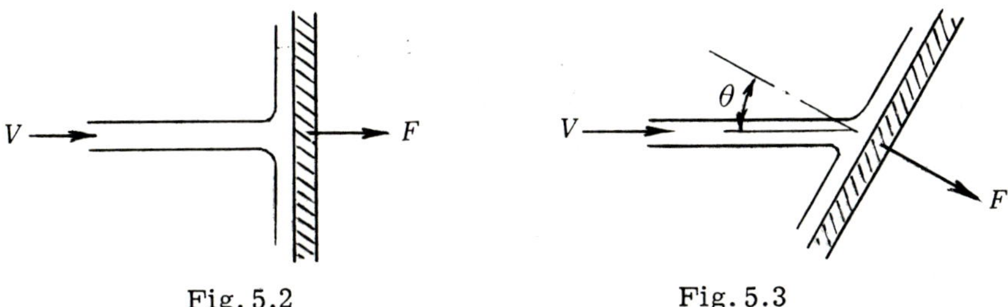

Fig. 5.2

Fig. 5.3

(*b*) *Flat plate inclined to jet* If the normal to the plate is inclined at an angle θ to the axis of the jet, Fig. 5.3, the initial velocity normal to the plate is $V\cos\theta$. The final velocity normal to the plate is again zero, so that

$$F = \dot{m}(V\cos\theta - 0)$$

$$= \rho a V^2 \cos\theta \qquad (5.2)$$

(*c*) *Symmetrical curved vane* Fig. 5.4 shows a curved vane of the type used in a Pelton wheel. If there is no friction as the liquid passes over the vane, the magnitude of the velocity remains unchanged. The initial velocity normal to the vane is V and the component of its final velocity normal to the vane is $-V\cos\theta$.

Hence $$F = \dot{m}(V - [-V\cos\theta])$$

$$= \rho a V^2(1 + \cos\theta) \qquad (5.3)$$

If the jet is completely reversed in direction, i.e. $\theta = 0$, then

$$F = 2\rho a V^2$$

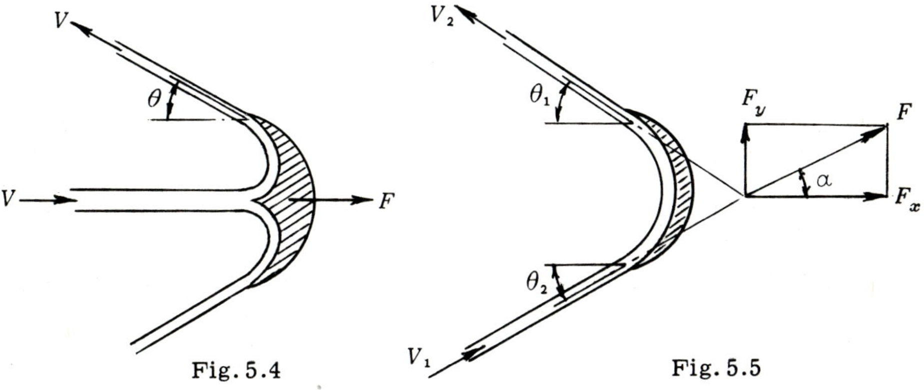

Fig. 5.4

Fig. 5.5

(*d*) *General case of curved vane* If the inlet and outlet velocities are V_1 and V_2 respectively, Fig. 5.5, then in the horizontal direction,

$$F_x = \dot{m}(V_1\cos\theta_1 + V_2\cos\theta_2) \qquad (5.4)$$

and in the vertical direction,

$$F_y = \dot{m}(V_1 \sin\theta_1 - V_2 \sin\theta_2) \quad (5.5)$$

The resultant force on the vane is given by

$$F = \sqrt{F_x^2 + F_y^2}$$

and the inclination to the horizontal is given by

$$\alpha = \tan^{-1}\frac{F_y}{F_x}$$

5.3 Impact of jets on moving plates If the plate shown in Fig. 5.6 is moving with velocity v in the direction of the jet, the mass flow rate of liquid impinging on the plate is given by

$$\dot{m} = \rho a(V - v)$$

The velocity of the jet is reduced from V to v after striking the plate, so that the change of momentum per second is $\dot{m}(V - v)$.

Thus $F = \dot{m}(V - v) = \rho a(V - v)^2$ (5.6)

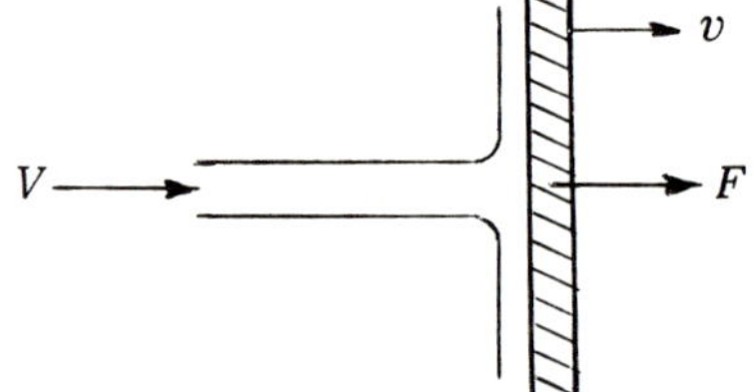

Fig. 5.6

This is an impracticable case since the jet would need to be continually lengthened.

A practical example of impact on moving plates is the water wheel, Fig. 5.7, where a series of flat plates is mounted radially on the rim of the wheel.

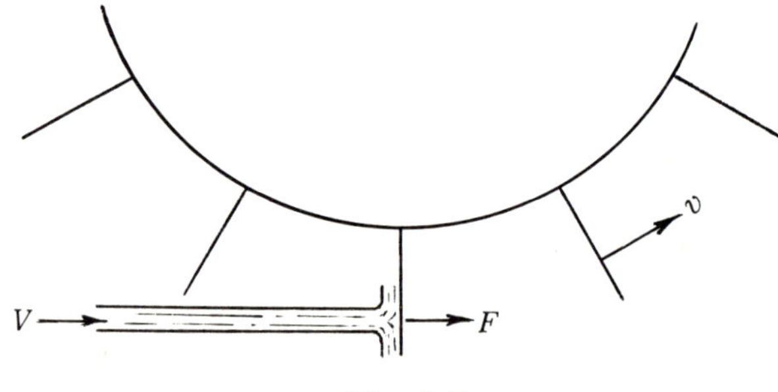

Fig. 5.7

The whole of the water leaving the nozzle strikes the plates, although the change in velocity is still $(V - v)$.

Hence $F = \dot{m}(V - v)$ where $\dot{m} = \rho a v$

Work done per second, or power output $= \dot{m}(V - v)v$ (5.7)

Energy supplied per second, or power input $= \frac{1}{2}\dot{m}V^2$

$$\text{Therefore efficiency, } \eta = \frac{\dot{m}(V - v)v}{\tfrac{1}{2}\dot{m}V^2}$$

$$= \frac{2(V - v)v}{V^2} \tag{5.8}$$

For maximum efficiency, $\dfrac{d\eta}{dv} = 0$ from which $v = \dfrac{V}{2}$

$$\text{Therefore maximum efficiency} = \frac{2\left(V - \frac{V}{2}\right)\frac{V}{2}}{V^2}$$

$$= 0{\cdot}5 \quad \text{or} \quad 50\%$$

5.4 Curved vanes and the Pelton wheel If the flat plates of a water wheel are replaced by curved vanes, shown in plan view in Fig. 5.8(*a*), the relative velocity of the jet to the vane at inlet is $(V - v)$ and, in the absence of friction, this will also be the velocity of the liquid relative to the vane at outlet.

Hence the absolute velocity of the liquid in the direction of motion of the vane is $v - (V - v)\cos\theta$, Fig. 5.8(*b*).

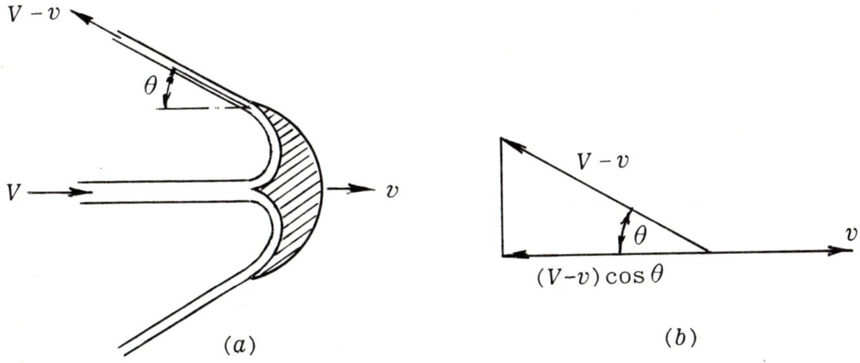

Fig. 5.8

$$\text{Thus force on vane, } F = \dot{m}\{V - [v - (V - v)\cos\theta]\}$$

$$= \dot{m}(V - v)(1 + \cos\theta)$$

$$\text{Power output} = \dot{m}(V - v)(1 + \cos\theta)v \tag{5.9}$$

$$\text{Power input} = \tfrac{1}{2}\dot{m}V^2$$

$$\text{Hence} \quad \eta = \frac{2(V - v)(1 + \cos\theta)v}{V^2} \tag{5.10}$$

For maximum efficiency, $v = \dfrac{V}{2}$ as before and the maximum efficiency becomes

$$\eta_{max} = \tfrac{1}{2}(1 + \cos\theta)$$

If $\theta = 0$, i.e. the jet is reversed in direction, the maximum efficiency is 100%, neglecting friction.

The above case is the basis of the Pelton wheel, Fig. 5.9. The vertical wheel, W, carries a series of curved vanes, or buckets, B, attached radially to its circumference. The water enters through the nozzle, N, controlled by a needle-valve, and strikes the vanes in a direction tangential to the wheel one half being deflected to either side. The lower edge of the vane is cut away so as not to impede the jet before it has come into the vertical position, Fig. 5.10.

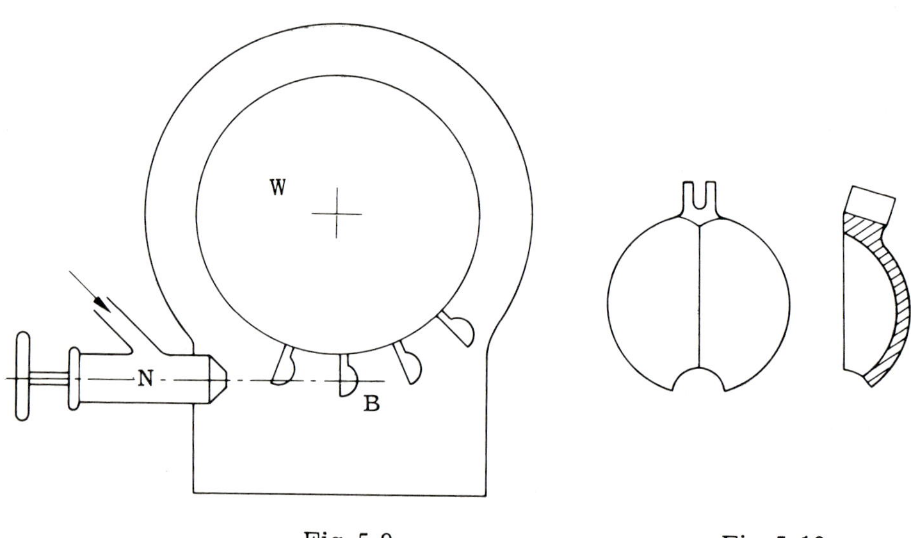

Fig. 5.9 Fig. 5.10

Due to friction, the relative velocity at outlet is slightly less than the relative velocity at inlet and the greatest angle through which the jet can be deflected is about 160°, otherwise the deflected jet will strike the back of the oncoming vane.

It is found that the maximum efficiency occurs when the vane speed is about 0·46 of the jet speed and the greatest efficiency available is about 85%.

5.5 Flow through turbines, fans, compressors and pumps The Pelton wheel is a simple form of impulse turbine. There are also impulse steam and gas turbines, reaction water, steam and gas turbines, fans, compressors and pumps, all of which use the principle that the rate of change of momentum of the working fluid involves forces which give power in the case of turbines and absorb power in the case of fans, compressors and pumps. In all these designs, the fluid is enclosed by a casing and the change of velocity as the fluid flows over the curved vanes is determined from relative velocity diagrams, (see Chapter 11).

The machines are usually of the axial flow or radial flow type, as shown in Figs. 5.11(a) and (b) respectively.

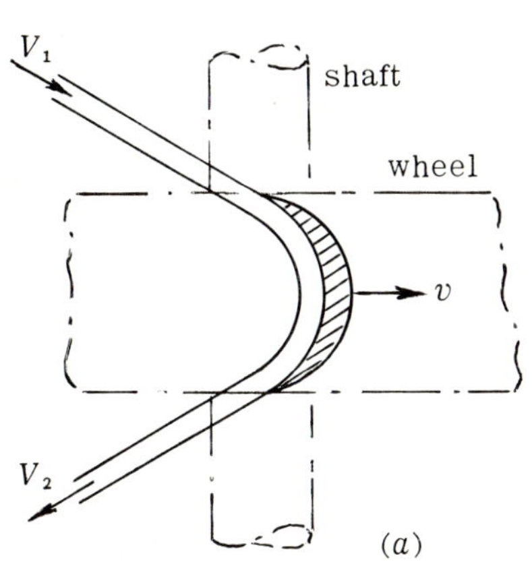

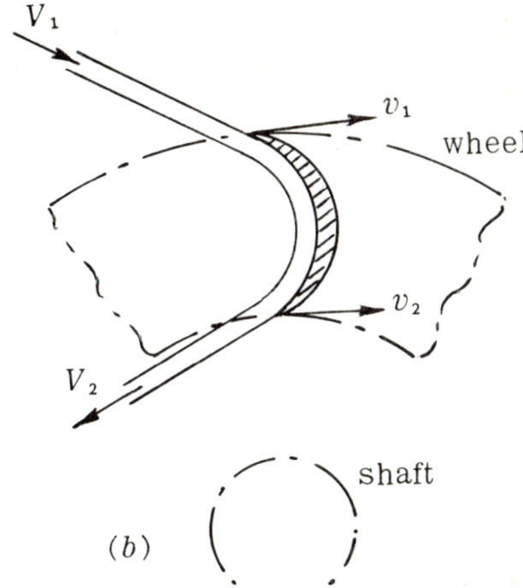

Fig. 5.11

For each type, the relative velocity diagrams are constructed at inlet and outlet to the vanes. Referring to Fig. 5.12, V_1 and V_2 are the absolute velocities of the fluid at inlet and outlet and v_1 and v_2 are the velocities of the vane tips. The relative velocities u_1 and u_2 are tangential to the vane tips so that the fluid enters and leaves the vane without shock.

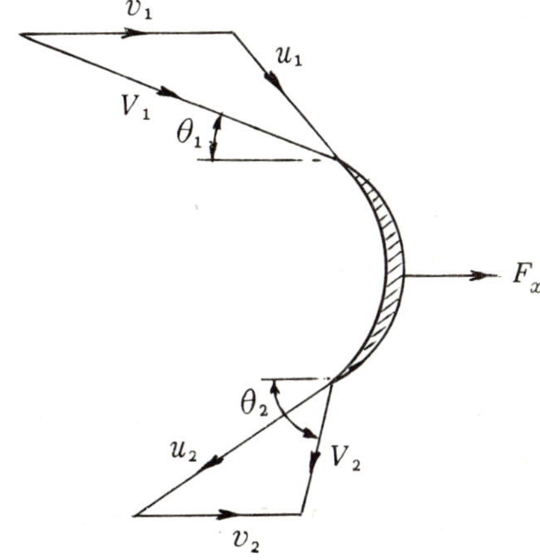

Fig. 5.12

The force on the vane,

$$F_x = \dot{m}(V_1 \cos\theta_1 + V_2 \cos\theta_2)$$

Torque on shaft,

$$T = \dot{m}(V_1 \cos\theta_1 r_1 + V_2 \cos\theta_2 r_2) \qquad (5.11)$$

where r_1 and r_2 are the radii of the vane tips at inlet and outlet respectively.

If ω is the angular velocity, then power output

$$= \dot{m}\omega(V_1 \cos\theta_1 r_1 + V_2 \cos\theta_2 r_2)$$

$$= \dot{m}(V_1 \cos\theta_1 v_1 + V_2 \cos\theta_2 v_2) \qquad (5.12)$$

In the case of an axial flow machine, $v_1 = v_2 = v$, so that

$$\text{power output} = \dot{m}v(V_1 \cos\theta_1 + V_2 \cos\theta_2) \qquad (5.13)$$

5.6 Forces on tapered pipes and bends Fig. 5.13 shows a tapered pipe in which the gauge pressure changes from p_1 to p_2 while the velocity changes from V_1 to V_2. The force exerted by the liquid on the pipe wall due to the change of momentum is $\dot{m}(V_1 - V_2)$, as before but there is an additional axial force $(p_1 A_1 - p_2 A_2)$ due to the pressure of the liquid.

Thus total force on pipe, $F = \dot{m}(V_1 - V_2) + (p_1 A_1 - p_2 A_2)$ (5.14)

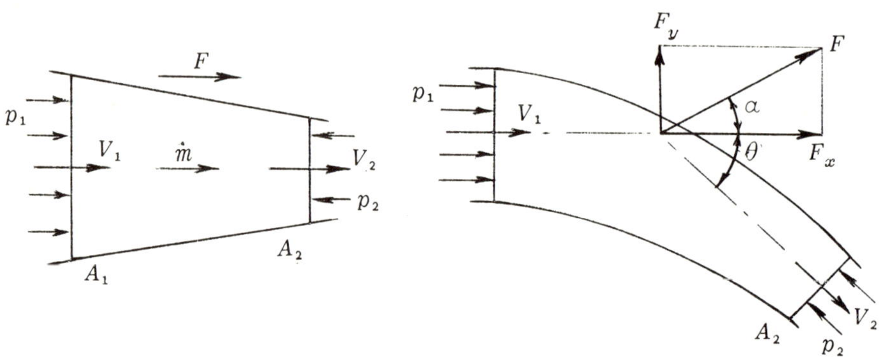

Fig. 5.13 Fig. 5.14

If the pipe turns through an angle θ between inlet and outlet, as shown in Fig. 5.14, then horizontal force on bend,

$$F_x = \dot{m}(V_1 - V_2\cos\theta) + (p_1 A_1 - p_2 A_2 \cos\theta)$$

and vertical force on bend,

$$F_y = \dot{m}(0 + V_2 \sin\theta) + (0 + p_2 A_2 \sin\theta)$$

The resultant force, $F = \sqrt{F_x^2 + F_y^2}$

and $\alpha = \tan^{-1}\dfrac{F_y}{F_x}$

5.7 Jet and propeller propulsion The aircraft and boat shown in Fig. 5.15 move with velocity v while the fluid leaves with velocity V relative to the craft.

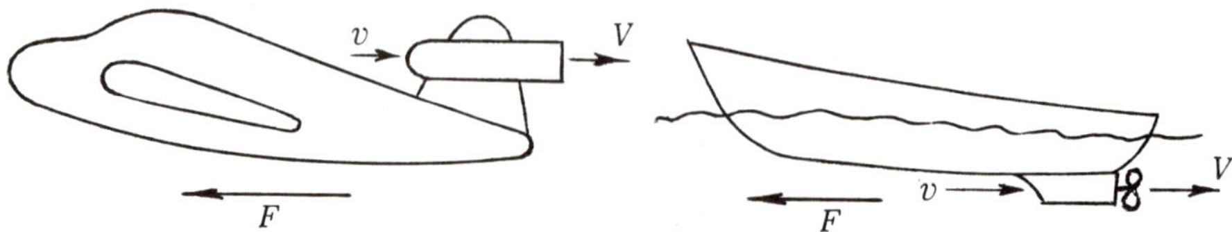

Fig. 5.15

The change of momentum of the fluid is $\dot{m}(V - v)$ and hence the propelling force,

$$F = \dot{m}(V - v)$$

where $\dot{m}$ is the mass of fluid ejected per second.

The propulsive efficiency is defined by

$$\eta = \frac{\text{work done on craft}}{\text{K.E. supplied to propellant}}$$

$$= \frac{Fv}{\frac{1}{2}\dot{m}(V^2 - v^2)}$$

$$= \frac{\dot{m}(V - v)v}{\frac{1}{2}\dot{m}(V^2 - v^2)} = \frac{2v}{V + v}$$

The efficiency rises as v rises but the thrust decreases and hence a compromise is needed.

1. *A jet of water flows smoothly on to a stationary curved vane and turns through 150°. The jet is 25 mm in diameter and is at atmospheric pressure. The jet leaves the vane with 80% of its entry velocity. If the flow rate is 0·01 m³/s, determine the force on the vane.*

The vane is shown in Fig. 5.16.

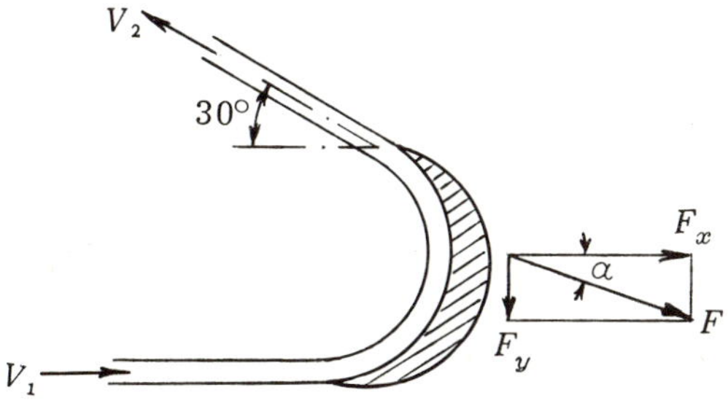

Fig. 5.16

$$\text{Mass flow rate} = 0{\cdot}01 \times 10^3 = 10 \text{ kg/s}$$

$$\text{Velocity of water entering vane, } V_1 = \frac{0{\cdot}01}{\frac{\pi}{4} \times \frac{25^2}{10^6}} = 20{\cdot}4 \text{ m/s}$$

$$\text{Velocity of water leaving vane, } V_2 = 0{\cdot}8 \times 20{\cdot}4 = 16{\cdot}3 \text{ m/s}$$

In the horizontal direction, change of momentum per second

$$= \dot{m}(V_1 - [-V_2 \cos 30°]) = \dot{m}(V_1 + V_2 \cos 30°)$$

i.e. $$F_x = 10(20{\cdot}4 + 16{\cdot}3 \times 0{\cdot}866) = 345 \text{ N}$$

In the vertical direction, change of momentum per second

$$= \dot{m}(0 + V_2 \sin 30^\circ)$$

i.e. $$F_y = 10 \times 16{\cdot}3 \times 0{\cdot}5 = 81{\cdot}5 \text{ N}$$

The resultant force, $F = \sqrt{345^2 + 81{\cdot}5^2} = \underline{354 \text{ N}}$

and the angle of the resultant to the horizontal is given by

$$\alpha = \tan^{-1}\frac{81{\cdot}5}{345} = \underline{13{\cdot}3^\circ}$$

2. *Water is supplied to a Pelton wheel through a nozzle 25 mm diameter with velocity 60 m/s. The buckets move in a path of mean radius 0·3 m and deflect the water through 160°. Calculate the speed of the wheel for maximum power and the value of this power (a) if the motion of the water over the buckets is assumed to be frictionless and (b) if friction reduces the velocity of the water as it passes over the vane by 10%.*

(*a*) For maximum efficiency, $v = \frac{1}{2}V = 30$ m/s . . . from Art. 5.4

$$\therefore \omega = \frac{v}{R} = \frac{30}{0{\cdot}3} = 100 \text{ rad/s}$$

$$\therefore N = \frac{60}{2\pi} \times 100 = \underline{955 \text{ rev/min}}$$

Maximum power $= \dot{m}(V - v)(1 + \cos\theta)v$ . . from equation (5.9)

$$= 10^3 \times \frac{\pi}{4} \times 0{\cdot}025^2 \times 60\,(60 - 30)(1 + \cos 20^\circ) \times 30$$

$$= 51\,400 \text{ W} \quad \text{or} \quad \underline{51{\cdot}4 \text{ kW}}$$

(*b*) If the exit velocity is reduced by 10%,

final momentum in original direction of jet $= -\rho a V \times 0{\cdot}9 V \cos\theta$

$$\therefore F = \rho a V^2 (1 + 0{\cdot}9\cos\theta)$$

The presence of the 0·9 in the formula for efficiency will not affect the condition for maximum efficiency, so that the speed for maximum efficiency is <u>955 rev/min</u>, as before.

The power will be reduced in the ratio $\dfrac{1 + 0{\cdot}9\cos\theta}{1 + \cos\theta}$

so that maximum power $= \dfrac{1 + 0{\cdot}9\cos 20^\circ}{1 + \cos 20^\circ} \times 51{\cdot}4$

$$= \underline{48{\cdot}8 \text{ kW}}$$

3. *An impulse steam turbine has blades with equal inlet and outlet angles and the pressure across the blade is constant. Steam enters with a velocity of 270 m/s, inclined at 11° to the direction of motion of the blades, and leaves with 90% of the relative velocity at which it entered. Calculate the mass flow rate of steam if the turbine is to develop 300 kW at a blade speed of 100 m/s.*

Fig. 5.17 shows the blade with the velocity triangles at inlet and outlet.

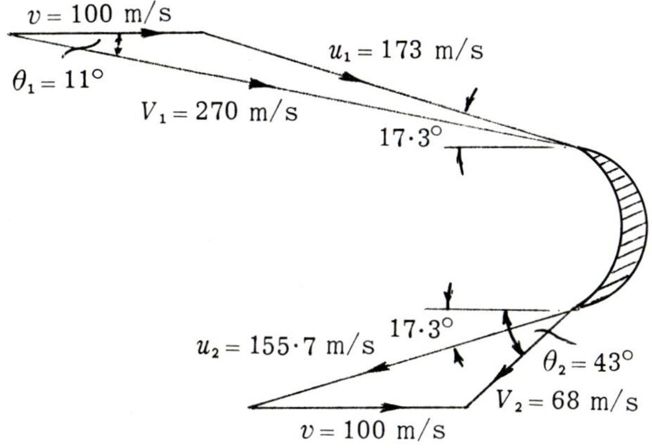

Fig. 5.17

From the geometry of the inlet triangle, the velocity of the steam relative to the blade,

$$u_1 = 173 \text{ m/s}$$

and the angle it makes with the direction of motion is 17·3°.

The relative velocity of the steam at outlet,

$$u_2 = 0{\cdot}9 \times 173 = 155{\cdot}7 \text{ m/s}$$

Due to the symmetry of the blade, this also makes an angle of 17·3° with the direction of motion.

From the outlet triangle, $V_2 = 68$ m/s

and $\theta_2 = 43°$

Force on blade in direction of motion,

$$F = \dot{m}(V_1 \cos\theta_1 + V_2 \cos\theta_2)$$

$$= \dot{m}(270 \cos 11° + 68 \cos 43°) = 314\,\dot{m}$$

$$\text{Power} = \text{force} \times \text{velocity}$$

i.e. $$300 \times 10^3 = 314\,\dot{m} \times 100$$

$$\therefore\ \dot{m} = \underline{9{\cdot}56 \text{ kg/s}}$$

4. *Water at a constant gauge pressure of 70 kN/m² flows through a pipe at 3 m/s. The pipe, which is 200 mm diameter, makes a 30° bend. Determine the force on the bend.*

$$\text{Mass flow rate, } \dot{m} = \rho a V$$

$$= 10^3 \times \frac{\pi}{4} \times 0{\cdot}2^2 \times 3 = 94{\cdot}2 \text{ kg/s}$$

Let F_x and F_y be the horizontal and vertical components of the force on the bend, Fig. 5.18(*a*).

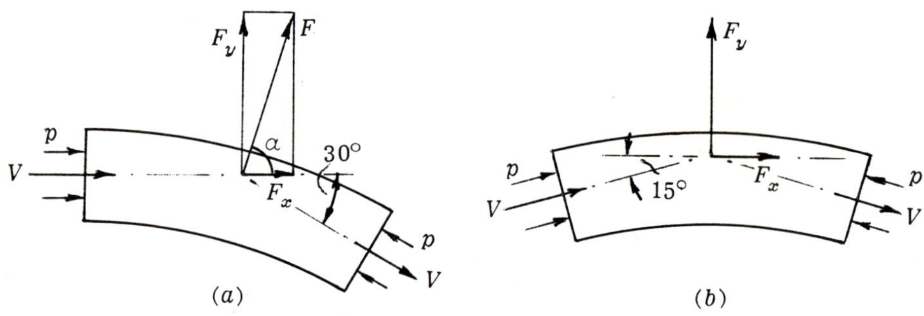

Fig. 5.18

Then, as in Art. 5.6,

$$F_x = \dot{m}(V - V\cos 30°) + (pA - pA\cos 30°)$$

$$= 94{\cdot}2 \times 3(1 - 0{\cdot}866) + 70 \times 10^3 \times \frac{\pi}{4} \times 0{\cdot}2^2(1 - 0{\cdot}866) = 323 \text{ N}$$

$$F_y = \dot{m}(0 + V\sin 30°) + (0 + pA\sin 30°)$$

$$= 94{\cdot}2 \times 3 \times 0{\cdot}5 + 70 \times 10^3 \times \frac{\pi}{4} \times 0{\cdot}2^2 \times 0{\cdot}5 = 1\,240 \text{ N}$$

$$\therefore F = \sqrt{323^2 + 1\,240^2} = \underline{1\,280 \text{ N}}$$

and $$\alpha = \tan^{-1}\frac{1\,240}{323} = \underline{75°}$$

Alternatively, due to the symmetry of the bend, Fig. 5.18(*b*),

$$F_x = \dot{m}(V\cos 15° - V\cos 15°) + (pA\cos 15° - pA\cos 15°)$$

$$= 0$$

and $$F_y = \dot{m}(V\sin 15° + V\sin 15°) + (pA\sin 15° + pA\sin 15°)$$

$$= 2 \times 94{\cdot}2 \times 3 \times 0{\cdot}259 + 2 \times 70 \times 10^3 \times \frac{\pi}{4} \times 0{\cdot}2^2 \times 0{\cdot}259$$

$$= \underline{1\,280 \text{ N}}$$

This is directed along the axis of symmetry.

5. *A speedboat is propelled by a circular jet of fresh water. The water enters the system through a forward-facing intake and is pumped horizontally through the stern. At 8 m/s the resistance of the boat is 950 N and the power supply to the pump is 14 kW. The pump efficiency is 90%.*

Ignoring the difference in elevation between the intake and the jet, determine the mass flow rate of water, the jet diameter and the efficiency of the propulsion system.

Let $\dot{m}$ be the mass flow rate, v be the velocity of the boat and V be the velocity of the jet relative to the boat, Fig. 5.19.

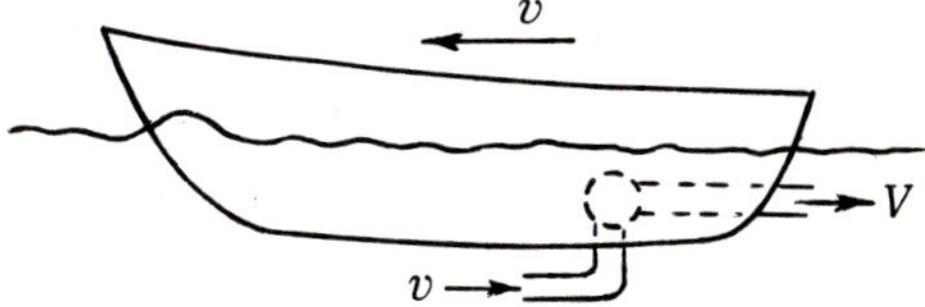

Fig. 5.19

$$\text{Power supplied to water} = 14 \times 0{\cdot}9 = 12{\cdot}6 \text{ kW}$$

$$\text{Increase in K.E. of water per second} = \tfrac{1}{2}\dot{m}(V^2 - v^2)$$

$$\therefore \quad 12{\cdot}6 \times 10^3 = \tfrac{1}{2}\dot{m}(V^2 - 8^2)$$

or

$$25\,200 = \dot{m}(V^2 - 64) \qquad (1)$$

$$\text{Propelling force on boat} = \dot{m}(V - v)$$

i.e.

$$950 = \dot{m}(V - 8) \qquad (2)$$

From equations (1) and (2),

$$\frac{25\,200}{950} = \frac{V^2 - 64}{V - 8} = V + 8$$

$$\therefore \quad V = 18{\cdot}5 \text{ m/s}$$

and

$$\dot{m} = \frac{950}{18{\cdot}5 - 8} = \underline{90{\cdot}5 \text{ kg/s}}$$

If the jet diameter is d, then

$$\dot{m} = \rho a V = \rho \frac{\pi}{4} d^2 V$$

i.e.

$$90{\cdot}5 = 10^3 \times \frac{\pi}{4} d^2 \times 18{\cdot}5$$

from which

$$d = \underline{0{\cdot}079 \text{ m}}$$

$$\text{Propulsion efficiency} = \frac{\text{output power}}{\text{input power}} = \frac{\text{thrust} \times \text{velocity}}{\text{energy supply to pump}}$$

$$= \frac{950 \times 8}{14 \times 10^3} = 0{\cdot}543 \quad \text{or} \quad \underline{54{\cdot}3\,\%}$$

6. A jet of water 25 mm in diameter and having a velocity of 8 m/s strikes a flat plate. Calculate the force on the plate (*a*) if it is stationary, (*b*) if it moves in the same direction as the jet at 3 m/s. (*Ans.*: 31·42 N; 12·28 N)

7. A 20 m/s wind blows against the side of a tanker in ballast. The tanker is 250 m long and has an average freeboard of 12 m. The density of air is 1·2 kg/m³. Calculate the normal force on the ship when the wind is abeam (i.e. at 90° to the ship's axis). (*Ans.*: 1·44 MN)

8. A jet of water issuing from a nozzle under a pressure head of 40 m strikes normally against a fixed flat vane and exerts on it a force of 800 N. What must be the diameter of the nozzle?

If, instead of a single fixed vane, there were a series of flat vanes moving in the direction of the jet at a speed of 12 m/s, each normal to the jet, what force would then be exerted on the vanes and at what rate would work be done on the vanes by the jet? Comparing the work done with the energy of the jet, what efficiency does this represent? (*Ans.*: 36 mm; 457·5 N; 5·49 kW; 49%)

9. A jet of water, 50 mm diameter, issues from a nozzle under a head of 60 m and strikes a series of moving vanes, each of which deflects the jet through an angle of 150°. If the velocity of the vanes in the direction of the jet is 10 m/s, calculate (*a*) the force acting on each vane, (*b*) the work done per second, (*c*) the efficiency. (*Ans.*: 3·055 kN; 30·55 kW; 77%)

10. A jet of water 80 mm diameter, moving with a velocity of 25 m/s, flows horizontally on to a stationary curved vane which deflects it through 120°. Calculate the magnitude and direction of the resultant force on the vane.

If the jet impinges on a series of such vanes moving at 12 m/s in the direction of the jet, determine the force on the vanes in the direction of the jet. At what speed must the vanes move for maximum work transfer rate?
(*Ans.*: 5·45 kN at 60° to jet; 3·19 kN; 12·5 m/s)

11. A Pelton wheel works under an effective head of 700 m. The jet is 150 mm diameter, and the efficiency is 89%. The jet is deflected through an angle of 155°. Calculate the quantity of water discharged, the power and the speed in rev/min if the mean diameter of the wheel is 1 m.
(*Ans.*: 2·07 m^3/s; 12·66 MW; 834 or 1 410 rev/min)

12. The buckets of a Pelton wheel deflect a jet, having a velocity of 60 m/s,, through an angle of 160°. Assuming that the velocity of the jet relative to the buckets is reduced by 15% as the jet water moves over them, find the efficiency of the wheel if the speed of the buckets is 27 m/s and, using this efficiency, calculate the diameter of the jet so that the wheel may develop 200 kW.
(*Ans.*: 89%; 51·5 mm)

13. A Pelton wheel water turbine has blades shaped as those in Fig. 5.8(*a*). The blade speed is 14 m/s and the turbine receives 0·7 m^3/s of water from a reservoir with a water level 30 m above the jet. The blades deflect the jet through 160°. Neglecting friction in the pipe, determine the power output and the efficiency of the turbine. (*Ans.*: 193·5 kW; 94%)

14. Water enters the blades of an impulse turbine with a velocity of 70 m/s in a direction inclined at 20° to the direction of motion of the blades. The velocity of the blades is 27 m/s and they are symmetrical in section, i.e., the inlet and outlet angles are equal. When the mass flow rate of water is 150 kg/s, determine the force on the blades and the power output. (*Ans.*: 11·6 kN; 313 kW)

15. A 90° horizontal bend reduces a pipe from 200 mm to 100 mm in diameter. The gauge pressure at entry to the 200 mm end is 300 kN/m^2. If the flow rate through the pipe is 0·1 m^3/s, determine the magnitude of the force on the bend.
(*Ans.*: 10·2 kN)

16. A horizontal bend reduces a pipe from 150 mm diameter to 75 mm diameter and deflects the axis through 60°. The gauge pressure in the 150 mm diameter pipe is 350 kN/m^2 and the flow is 0·085 m^3/s. Find the forces on the bend parallel and normal to the 150 mm diameter pipe. (*Ans.*: 5·386 kN; 2·092 kN)

17. The resistance to motion of a vessel in fresh water is 22 kN at 5 m/s. The vessel is propelled by water jet; the water enters through a forward-facing intake at the bow and is discharged at the stern. The propulsive efficiency is 0·8 and the mechanical efficiency of the jet pumps is 0·75. Determine the jet velocity, the jet exit area and the power required to drive the pumps.
(*Ans.*: 2·5 m/s; 1·175 m^2; 183·5 kW)

6 Viscosity

6.1 Introduction The resistance to flow of a fluid is due to molecular cohesion, which results in a shearing action as layers of the fluid slide relative to each other. This resistance to shear stress is a measure of the *viscosity* of the fluid; simple experiments will show that viscosity is different for different fluids and that it varies with temperature. Thus oil is more viscous than water and its viscosity falls as the temperature rises.

6.2 Dynamic and kinematic viscosity When a fluid flows smoothly over a stationary boundary, Fig. 6.1, the layer in contact with the boundary is at rest and subsequent layers move with increasing velocities as the distance from the boundary increases. Thus there is a velocity gradient across the section of flow.

There is a shearing action between adjacent layers resulting in a loss of energy as the resisting force is overcome. If the velocity changes by dv in a distance dy perpendicular to the direction of flow, Fig. 6.1, the viscous strain rate, $\phi = dv/dy$.

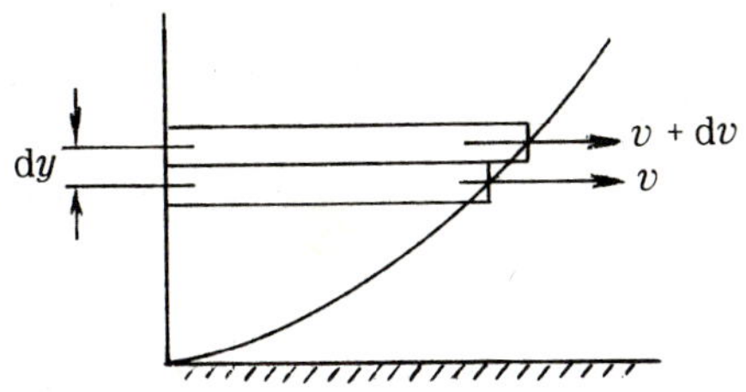

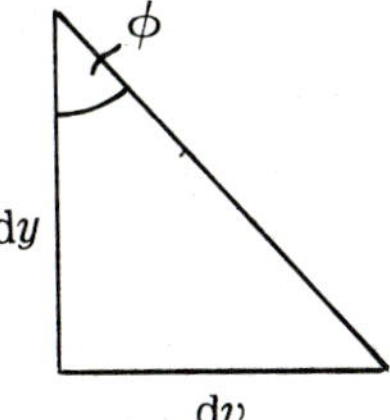

Fig. 6.1

The viscous stress τ is the viscous resistance per unit area and the coefficient of viscosity μ is defined as the ratio viscous stress/viscous strain rate,

i.e.
$$\mu = \frac{\tau}{\phi} = \frac{\tau}{dv/dy}$$

or
$$\tau = \mu \frac{dv}{dy} \tag{6.1}$$

The coefficient of viscosity is also called the *dynamic viscosity* to distinguish it from the *kinematic viscosity* ν, defined by

$$\nu = \frac{\mu}{\rho} \tag{6.2}$$

Kinematic viscosity is not a fundamental property of a fluid but the ratio μ/ρ occurs frequently in problems of fluid motion and it is found convenient to give it a separate name.

The units of μ are kg/m s (1 poise = 0·1 kg/m s)

and the units of ν are m^2/s (1 stoke = 10^{-4} m^2/s) .

6.3 Measurement of viscosity Viscosity is usually determined by a *viscometer*, which measures the time taken for a fixed volume of liquid to flow through a capillary tube. A calibration table then relates the time in seconds to the viscosity of the liquid.

6.4 Redwood viscometer Fig. 6.2 shows the Redwood viscometer, frequently used in Britain. The liquid under test is contained in a beaker with a capillary tube in the base. The liquid temperature is controlled by a water bath which is heated and stirred. The time taken for 50 ml of liquid to pass through the tube is then observed.

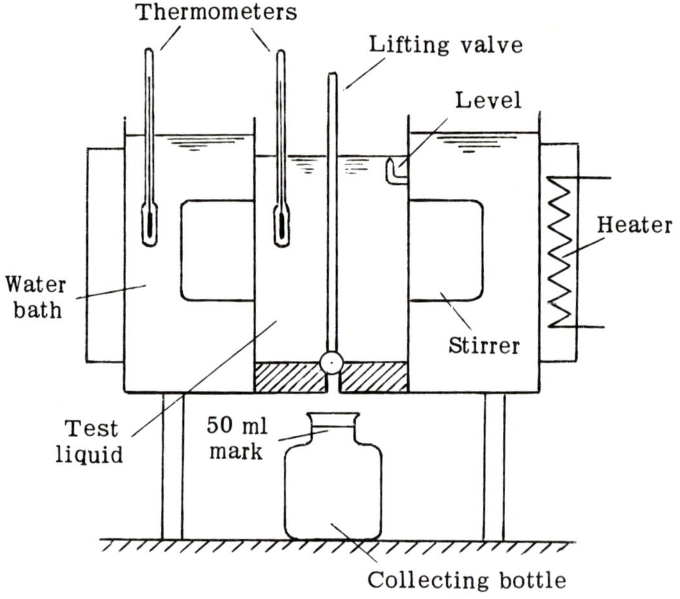

Fig. 6.2

For viscous flow through a round tube of diameter d and length l, the velocity of flow is given by

$$h_f = \frac{32\mu l V}{\rho g d^2} \quad . \quad . \quad . \quad . \quad \text{from equation (7.5)}$$

where h_f is the head loss due to viscous friction.

Thus quantity flowing,

$$Q = \frac{\pi}{4}d^2 \times V = \frac{\pi}{4}d^2 \times \frac{h_f \rho g d^2}{3\mu l}$$

or

$$Q \propto \frac{\rho}{\mu} \propto \frac{1}{\nu}$$

If t is the time for a fixed volume of liquid to pass, then $t \propto \frac{1}{Q}$, so that

$$t \propto \nu \qquad (6.3)$$

This equation is not quite correct since part of the available head is used to produce velocity rather than overcome friction losses in the tube, so that a more correct equation is

$$\nu = At - \frac{B}{t}, \quad \text{where } A \text{ and } B \text{ are constants.}$$

This equation is embodied in the conversion table provided with the equipment, relating time (Redwood seconds) with kinematic viscosity.

The Redwood No. 1 viscometer is suitable for viscosities less than 50×10^{-4} m²/s and the Redwood No. 2 viscometer, having a larger diameter exit tube, is used for more viscous liquids. The time from the No. 2 viscometer is multiplied by 10 to convert to Redwood (No. 1) seconds.

6.5 British Standard U-tube viscometer In the U-tube viscometer, Fig. 6.3, the liquid under test flows from a reservoir through a capillary tube. The instrument is initially filled to the filling mark and placed vertically in a constant temperature water bath. The liquid is then forced by pressure to the top of the other limb and allowed to flow back through the capillary tube, being timed between the two marks shown. The kinematic viscosity is again proportional to the time taken to flow through the capillary tube,

$$\nu = Ct \tag{6.4}$$

The value of the constant C is supplied with the instrument.

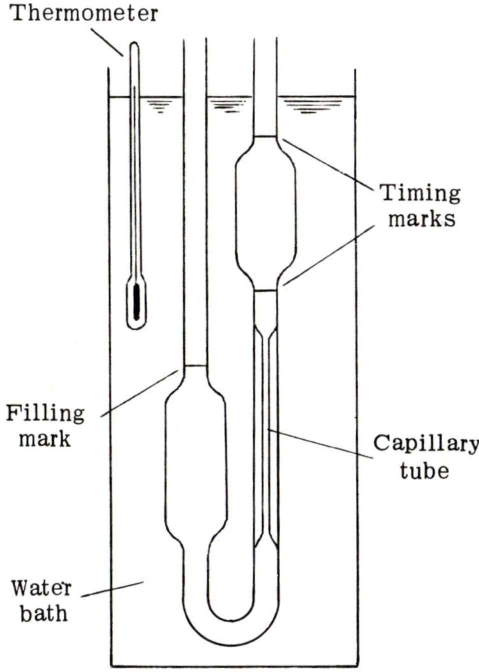

Fig. 6.3

6.6 Stoke's Law and terminal velocity When a small sphere of diameter d falls with velocity v through a liquid of viscosity μ, the force R opposing the motion is given by *

$$R = 3\pi\mu v d \qquad (6.5)$$

This is known as Stoke's Law.

If the sphere is released from rest, the velocity theoretically never reaches a limiting value but it effectively reaches a terminal velocity V in an extremely short time. The motion is then governed by the equilibrium equation,

downward force (weight) = upward force (buoyancy) + resistance

i.e.
$$\frac{\pi}{6}d^3\rho_s g = \frac{\pi}{6}d^3\rho g + 3\pi\mu V d$$

where ρ_s and ρ are the densities of the sphere and liquid respectively.

Hence
$$\mu = \frac{(\rho_s - \rho)gd^2}{18V} \qquad (6.6)$$

The initial time taken to reach equilibrium conditions may be eliminated by measuring V from a point below the liquid surface.

6.7 Viscosity determination by falling ball In the falling ball viscometer, Fig. 6.4, the instrument is held vertically in a water bath and the velocity of the ball is determined by timing its motion between two marks. Equation (6.6) could then be used to determine the viscosity of the liquid but this assumes that the liquid surrounding the ball is of infinite extent.

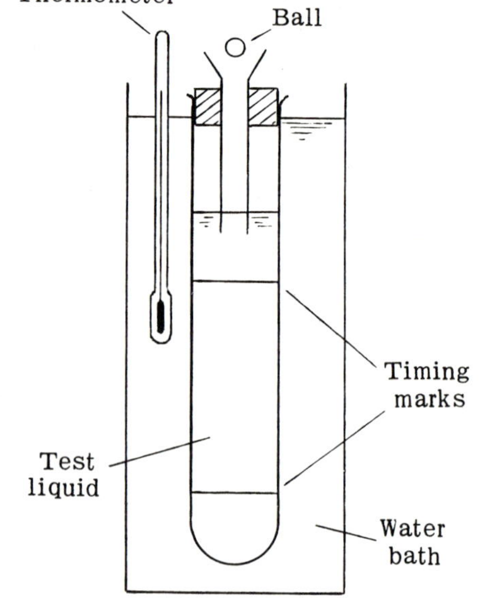

Fig. 6.4

Due to the disturbance caused by the viscometer tube, equation (6.6) is multiplied by a correction factor F which varies with the ratio d/D, where D is the tube diameter and d the ball diameter.

Thus
$$\mu = \frac{(\rho_s - \rho)gd^2}{18V} \times F$$

The ball size is chosen to ensure that $V < 1$ cm/s and that $d/D < 0{\cdot}1$. When $d/D = 0{\cdot}1$, F is approximately 0·8. Further details of viscosity measurement may be found in BS 188 : 1957.

* Provided that $Re < 0{\cdot}1$ (see Art. 7.4).

1. *A piston, 50 mm diameter and 75 mm long, moves vertically in an open-ended lubricated cylinder with a radial clearance of 0·1 mm. When falling due to its own weight, the piston moves through 30 mm in 4·2 s at a uniform velocity. When a mass of 0·05 kg is added to the piston, it moves with uniform velocity through the same distance in 2·4 s. Calculate the viscosity of the oil and the mass of the piston.*

$$\tau = \mu \frac{dv}{dy} \quad . \quad . \quad . \quad . \quad . \quad . \quad . \quad \text{from equation (6.1)}$$

If the mass of the piston is m, then

$$\tau = \frac{\text{axial force}}{\text{curved surface area of piston}}$$

$$= \frac{mg}{\pi \times 0{\cdot}05 \times 0{\cdot}075} = 84{\cdot}88\, mg \ \ \text{N/m}^2$$

$$dv = \frac{0{\cdot}03}{4{\cdot}2} = 0{\cdot}007\,143 \ \ \text{m/s}$$

$$dy = 0{\cdot}1 \times 10^{-3} \ \ \text{m}$$

$$\therefore \ 84{\cdot}88\, mg = \mu \times \frac{0{\cdot}007\,143}{0{\cdot}1 \times 10^{-3}}$$

$$\therefore \ mg = 0{\cdot}841\,5\,\mu \qquad (1)$$

When the additional 0·05 kg is added to the piston,

$$\tau = 84{\cdot}88\,(m + 0{\cdot}05)g$$

and

$$dv = \frac{0{\cdot}03}{2{\cdot}4} = 0{\cdot}012\,5 \ \ \text{m/s}$$

$$\therefore \ 84{\cdot}88\,(m + 0{\cdot}05)g = \mu \times \frac{0{\cdot}012\,5}{0{\cdot}1 \times 10^{-3}}$$

$$\therefore \ (m + 0{\cdot}05)g = 1{\cdot}473\,\mu \qquad (2)$$

Therefore, from equations (1) and (2),

$$\mu = \underline{0{\cdot}777 \ \ \text{kg/m s}}$$

and

$$m = \underline{0{\cdot}066\,6 \ \ \text{kg}}$$

2. *A steel sphere, 1·6 mm diameter, has a mass of 16·6 mg. The ball falls at its terminal velocity in oil of density 950 kg/m³, covering 300 mm in 34 s. Neglecting the wall effects of the enclosing vessel, determine the viscosity of the oil.*

$$V = \frac{0{\cdot}3}{34} = 8{\cdot}82 \times 10^{-3} \ \ \text{m/s}$$

Assuming that Stoke's Law is applicable,

$$\mu = \frac{(\rho_s - \rho)gd^2}{18V} \quad . \quad . \quad . \quad . \quad . \quad \text{from equation (6.6)}$$

$$\rho_s = \frac{\text{mass of sphere}}{\text{volume of sphere}}$$

$$= \frac{16{\cdot}6 \times 10^{-6}}{\frac{\pi}{6} \times (1{\cdot}6 \times 10^{-3})^3} = 7\,740 \text{ kg/m}^3$$

$$\therefore \mu = \frac{(7\,740 - 950) \times 9{\cdot}81 \times (1{\cdot}6 \times 10^{-3})^2}{18 \times 8{\cdot}82 \times 10^{-3}}$$

$$= \underline{1{\cdot}074 \text{ kg/m s}}$$

The validity of Stoke's Law may now be verified, i.e. that $Re < 0{\cdot}1$.

$$Re = \frac{\rho V d}{\mu} = \frac{950 \times 8{\cdot}82 \times 10^{-3} \times 1{\cdot}6 \times 10^{-3}}{1{\cdot}074}$$

$$= \underline{0{\cdot}0125}$$

3. A piston 100 mm diameter and 200 mm long is enclosed in an open-ended lubricated cylinder containing oil of viscosity 0·8 kg/m s. The radial clearance between the piston and cylinder is 0·2 mm.

Calculate (*a*) the axial force required to move the piston at a uniform speed of 0·5 m/s;

(*b*) the torque required to rotate the piston at a uniform speed of 2 rad/s.

(*Ans*.: 126 N; 1·26 N m)

4. Assuming that Reynolds' number is 0·1, determine the settling velocity of the largest spherical particles of density 2 000 kg/m³ which will fall from air of density 1·23 kg/m³ and kinematic viscosity 15×10^{-6} m²/s in accordance with Stokes' Law. What is the mass of the particles? ((*Ans*.: 0·059 m/s; $1{\cdot}7 \times 10^{-8}$ g)

5. Assuming Stokes' Law applies, find the time taken for a uniform suspension of spherical particles of density 3 000 kg/m³ and diameter 0·02 mm to clear from water of depth 3 m. The viscosity of the water is 0·013 poise.

Check the validity of the assumption. (*Ans*.: 2·485 h)

7 Real fluid flow

7.1 Introduction In Chapter 4, the flow rates predicted by simple theory were not in agreement with measured values and the discrepancies were corrected by experimentally determined coefficients of discharge. These differences are due to the effects of viscosity, pipe walls, fittings, etc., and it is now necessary to determine the losses due to these causes.

7.2 Steady, unsteady and quasi-steady flow If the velocity of flow at a point in a fluid stream does not vary with time, the flow is said to be *steady*. and if it does vary with time, the flow is *unsteady*. In some cases, such as the flow from a reservoir, the level falls very slowly as water is drawn off and the velocity in the discharge pipe may be treated as steady, although varying slightly with time. Such flow is termed *quasi-steady*.

7.3 Uniform and non-uniform flow If the velocity of flow is constant at various points in a region, the flow is *uniform* but if it varies from point to point, it is *non-uniform*. Non-uniformity may occur in the direction of flow, such as in a tapering pipe, or normal to the direction of flow, such as in the cross-section of a pipe or duct.

7.4 Reynolds' experiment If a duplicate of Reynolds' original experiment, Fig. 7.1(*a*), is performed, in which a coloured dye is fed into a moving stream of water, it is found that at low speeds, the dye stream remains in a straight line, Fig. 7.1(*b*), and the flow is described as *laminar* or *viscous*. However, at high speeds, the dye becomes mixed throughout the water, Fig. 7.1(*c*), and the flow is described as *turbulent*. The speed at which the flow changes from laminar to turbulent is not clearly defined and there is a transition zone between the two patterns.

It is found that the change of pattern depends upon the velocity of flow V, the tube diameter d and the kinematic viscosity ν, and the ratio $\frac{Vd}{\nu}$ (or $\frac{\rho V d}{\mu}$) is known as the *Reynolds' Number* (Re).

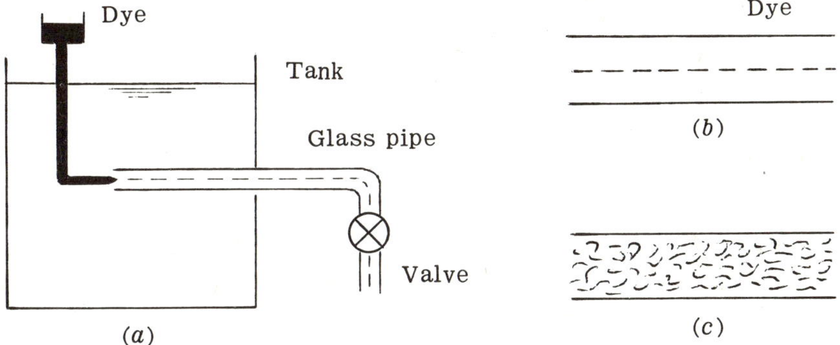

Fig. 7.1

It is also found that the transition occurs at slightly higher values of *Re* when the velocity of flow is increased than when it is decreased. These values are termed the *upper critical number* and *lower critical number* respectively; the lower value is considered the more important and is that defined as the *critical Reynolds' number*. In round pipes, flow is laminar for *Re* less than 2 000 and turbulent for *Re* greater than 4 000. The transition zone lying between these two values has flow which is unstable and indeterminate.

For liquids of low viscosity such as water, the velocity of flow at which the transition occurs is quite low and in most hydraulics problems, the flow will be turbulent.

7.5 Velocity distribution If a pitot tube is traversed across the diameter of a pipe conveying a moving liquid, it is found that the flow is non-uniform. Fig. 7.2 shows the velocity profiles for laminar and turbulent flow; for laminar flow, the profile is parabolic, as shown in Art. 7.8 but for turbulent flow, the velocity is almost uniform, except in the vicinity of the walls. In both cases, the velocity at the pipe wall is zero,and it is therefore impossible to have uniform flow adjacent to a boundary.

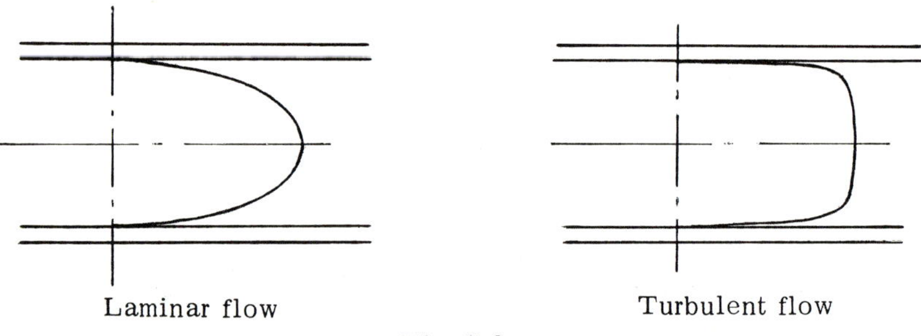

Fig. 7.2

To determine the flow rate for non-uniform velocity distribution, it is necessary to integrate the flow over small elements of the pipe section. For turbulent flow, the method was shown in Art. 4.8 but for laminar flow, the flow rate can be determined by mathematical integration, as in Art. 7.8.

7.6 Loss of head in pipes due to friction Reynolds performed further experiments to determine the loss of head in a pipe for both laminar and turbulent flow. The method is shown in Fig. 7.3 and it was found that the loss of head, h_f, over a length l increased as the velocity increased.

For laminar flow, the loss of head per unit length,

$$\frac{h_f}{l} \propto V$$

but for turbulent flow, the loss of head increased at a greater rate, given by

$$\frac{h_f}{l} \propto V^n$$

These results are shown in Fig. 7.4.

The friction loss for turbulent water flow had already been investigated by Darcy, who found that the loss of head was approximately proportional to V^2.

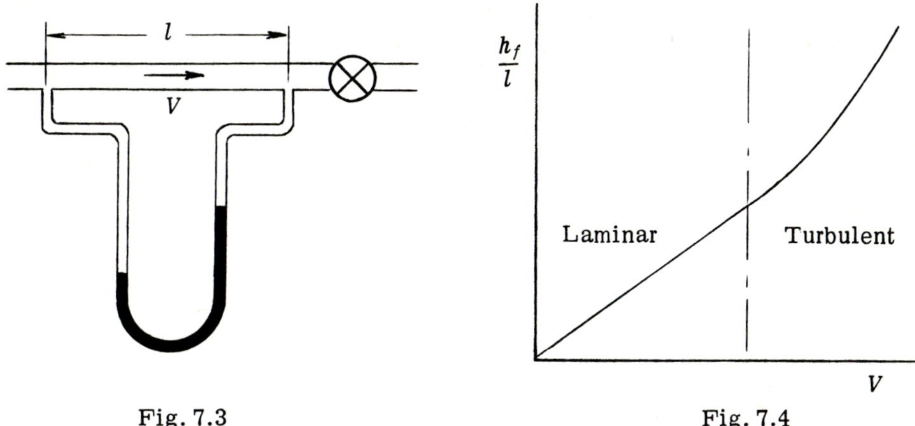

Fig. 7.3

Fig. 7.4

7.7 Darcy's formula Consider the flow of liquid through a pipe of diameter d and length l, Fig. 7.5, and let the viscous stress at the wall be τ.

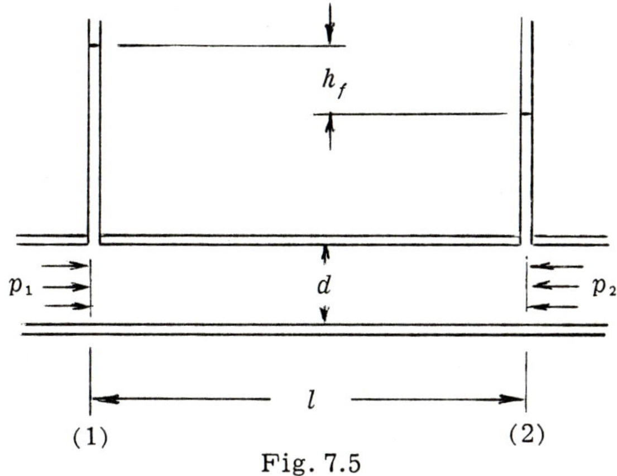

Fig. 7.5

Equating forces on the liquid between sections (1) and (2),

$$p_1 \times \frac{\pi}{4} d^2 - p_2 \times \frac{\pi}{4} d^2 = \tau \times \pi dl$$

Thus the loss of head due to friction,

$$h_f = \frac{p_1 - p_2}{\rho g} = \frac{4\tau l}{\rho g d}$$

A dimensionless friction coefficient f is defined by the relation $f = \frac{\tau}{\frac{1}{2}\rho V^2}$ where V is the average velocity in the pipe,

so that

$$h_f = \frac{4(\frac{1}{2}\rho V^2 f)l}{\rho g d} = \frac{4flV^2}{2gd} \qquad (7.1)$$

This is known as *Darcy's formula.*

The loss of head per unit length, $\frac{4f}{d}\cdot\frac{V^2}{2g}$, is termed the *hydraulic gradient* since it represents the gradient on which the pipe must be laid to maintain a uniform pressure. The hydraulic gradient is denoted by i, so that

$$h_f = il \tag{7.2}$$

7.8 Laminar flow through round pipes Fig. 7.6 shows the velocity distribution in a pipe of radius r and length l.

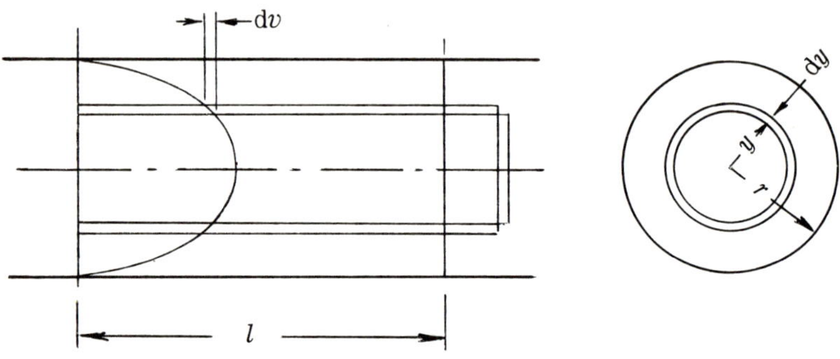

Fig. 7.6

When the radius increases from y to $y + \mathrm{d}y$, the velocity *decreases* by $\mathrm{d}v$, so that viscous strain rate $= -\frac{\mathrm{d}v}{\mathrm{d}y}$.

Viscous resistance of a cylinder of radius y and length l

$$= 2\pi yl \times \tau = 2\pi yl \times -\mu\frac{\mathrm{d}v}{\mathrm{d}y} \quad \text{from Art. 6.2.}$$

This resistance is equal to the difference in the forces on the ends of the cylinder,

i.e.
$$-2\pi yl\,\mu\frac{\mathrm{d}v}{\mathrm{d}y} = \rho g h_f \times \pi y^2 = \rho g i l \pi y^2$$

from which
$$\frac{\mathrm{d}v}{\mathrm{d}y} = -\frac{\rho g i}{2\mu}\cdot y$$

$$\therefore\ v = -\frac{\rho g i}{4\mu}y^2 + c$$

But $v = 0$ at $y = r$, so that
$$0 = -\frac{\rho g i}{4\mu}r^2 + c$$

from which
$$v = \frac{\rho g i}{4\mu}(r^2 - y^2) \tag{7.3}$$

Equation (7.3) shows that the velocity distribution in laminar flow is parabolic.

If V is the mean velocity, the flow rate is given by

$$Q = V \times \pi r^2 = \int_0^r \frac{\rho g i}{4\mu}(r^2 - y^2) \,.\, 2\pi y \, dy$$

$$= \frac{\pi \rho g i}{2\mu} \int_0^r (r^2 y - y^3) \, dy$$

$$= \frac{\pi \rho g i r^4}{8\mu} \tag{7.4}$$

Equation (7.4) is known as Poiseuille's equation

Thus $$V = \frac{\rho g i r^2}{8\mu} = \frac{\rho g i d^2}{32\mu}$$

$$\therefore \; h_f = il = \frac{32\mu l V}{\rho g d^2} \tag{7.5}$$

$$= \frac{32 l V^2}{g d \left(\frac{\rho V d}{\mu}\right)}$$

The term $\frac{\rho V d}{\mu}$ (or $\frac{V d}{\nu}$) is called the *Reynolds number* and is denoted by Re.

Thus $$h_f = \frac{32 l V^2}{g d (Re)}$$

Comparing this with the Darcy formula, equation (7.1),

$$h_f = \frac{4 f l V^2}{2 g d}$$

it will be seen that $$f = \frac{16}{Re} \quad \text{for laminar flow} \tag{7.6}$$

7.9 Turbulent flow through round pipes It is difficult to obtain a Reynolds number for water low enough to obtain laminar flow; thus the flow is generally turbulent, with a velocity profile as shown in Fig. 7.2. It is therefore impracticable to analyse the motion analytically, as for laminar flow, and the friction coefficient f in Darcy's formula is obtained experimentally.

It is found that for turbulent flow in *smooth* pipes,

$$f = \frac{0{\cdot}079}{(Re)^{1/4}} \tag{7.7}$$

and substituting this in the Darcy formula gives $h_f \propto V^{1{\cdot}75}$. However, it is usually sufficient to assume that $h_f \propto V^2$, i.e. that f is a constant.

For *rough* pipes, the surface texture affects the value of f and reference may be made to the Moody diagram, Fig. 7.7, which relates the values of f and Re for different degrees of roughness.

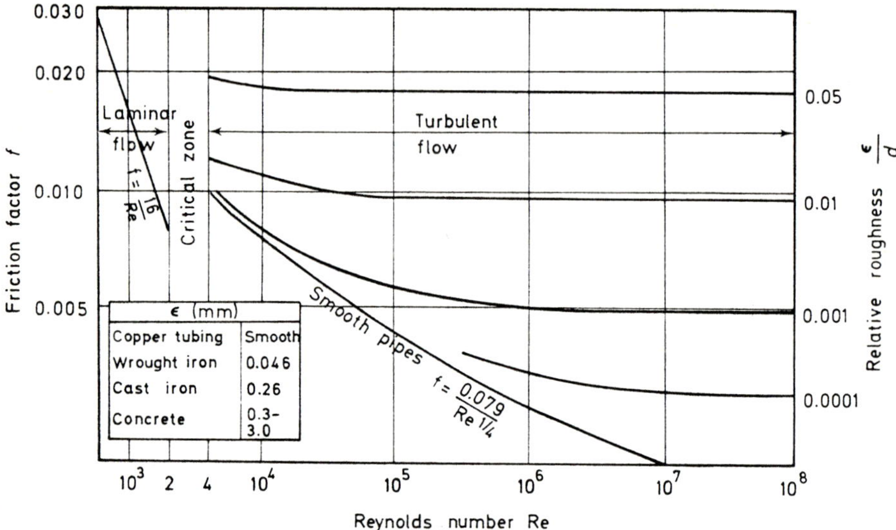

Fig. 7.7

7.10 Losses due to sudden enlargement and contraction in turbulent flow

(*a*) *Sudden enlargement* When the section of a pipe is suddenly enlarged, the pattern of flow is as indicated in Fig. 7.8. As the velocity head is reduced, the pressure at section (2) is higher than at section (1) but experiments show that the pressure immediately beyond the enlargement is the same as that at section (1).

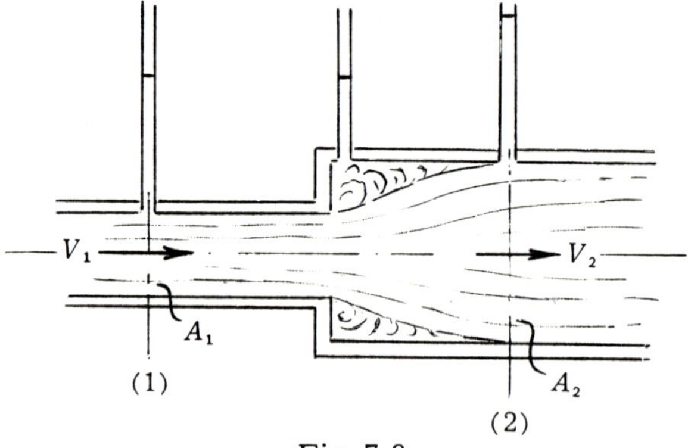

Fig. 7.8

Net force acting from left to right = change of momentum between sections (1) and (2)

i.e. $$p_1 A_1 + p_1(A_2 - A_1) - p_2 A_2 = \dot{m}(V_2 - V_1)$$

i.e. $$(p_1 - p_2)A_2 = \rho A_2 V_2 (V_2 - V_1)$$

or
$$\frac{p_1}{\rho g} - \frac{p_2}{\rho g} = \frac{V_2}{g}(V_2 - V_1)$$

If the loss of head between sections (1) and (2) is h_l, then from Bernoulli's equation,

$$\frac{p_1}{\rho g} + \frac{V_1^2}{2g} = \frac{p_2}{\rho g} + \frac{V_2^2}{2g} + h_l$$

Thus
$$\frac{p_1}{\rho g} - \frac{p_2}{\rho g} = \frac{V_2^2}{2g} - \frac{V_1^2}{2g} + h_l$$

$$\therefore \frac{V_2^2}{2g} - \frac{V_1^2}{2g} + h_l = \frac{V_2^2}{g} - \frac{V_1 V_2}{g}$$

$$\therefore h_l = \frac{V_1^2}{2g} - \frac{V_1 V_2}{g} + \frac{V_2^2}{2g}$$

$$= \frac{(V_1 - V_2)^2}{2g} \qquad (7.8)$$

Since $A_1 V_1 = A_2 V_2$, equation (7.8) may be expressed in the form

$$h_l = \left(\frac{A_2}{A_1} - 1\right)^2 \frac{V_2^2}{2g} \qquad (7.9)$$

(*b*) *Sudden contraction* The flow pattern in a sudden contraction is shown in Fig. 7.9, a vena contracta forming at section (1). The flow then expands to fill the pipe at section (2) and the only loss occurs in this expansion process. Thus, applying equation (7.9),

$$h_l = \frac{V_2^2}{2g}\left(\frac{A_2}{A_1} - 1\right)^2$$

$$= \frac{V_2^2}{2g}\left(\frac{1}{c_c} - 1\right)^2$$

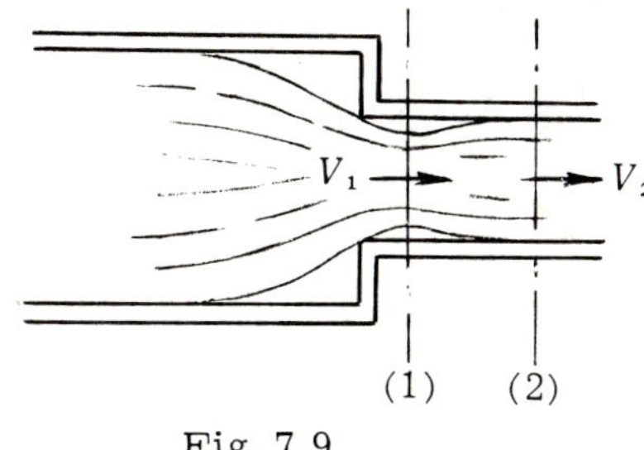

Fig. 7.9

where c_c is the coefficient of contraction.

By experiment, it is found that $\left(\frac{1}{c_c} - 1\right)^2$ is about 0·5, so that

$$h_l = 0{\cdot}5 \frac{V_2^2}{2g} \qquad (7.10)$$

An entry to a tank or exit from a pipe may be treated as an enlargement where A_2 is infinite, so that the loss is $V^2/2g$, i.e. all the kinetic energy at entry or exit is wasted.

The exit from a tank or entry to a pipe may be treated as a contraction, so that the loss is $0{\cdot}5\frac{V^2}{2g}$ where V is the exit or entry velocity.

7.11 Other losses in turbulent pipe flow In general, losses of head due to bends, valves, etc., are expressed as a fraction of the velocity head,

i.e. $$h_l = k\frac{V^2}{2g} \quad \text{where } k \text{ is a constant}$$

Typical values of k are as follows:

Smooth bend	0·30	Angle valve (open)	5·0
Mitre bend	1·1	Gate valve (open)	0·19
90° elbow	0·9	Gate valve (¾ open)	1·15
45° elbow	0·42	Gate valve (½ open)	5·6
Standard T	1·8	Conical enlargement (10°)	0·16
Globe valve (open)	10·0	Conical enlargement (25°)	0·55

(Data from Mechanical Engineer's Reference Book, Butterworth, 1973.)

In general terms, losses may be minimized by making changes in section or direction as gradual as possible. For example, a well-designed venturi meter involves practically no loss of head.

7.12 Total head and hydraulic gradient Fig. 7.10 shows a pipe line connecting two tanks or reservoirs. The total head at any section is represented by $Z + \frac{p}{\rho g} + \frac{V^2}{2g}$ and if this quantity is plotted along the pipe, it illustrates the distribution of the losses.

If the velocity head is subtracted from the total energy line, the remaining quantity $Z + \frac{p}{\rho g}$ represents the hydraulic gradient or pressure line. This shows the head available for distribution between pressure and elevation above datum and may be used to determine the pressure at points in a pipe line which varies in altitude.

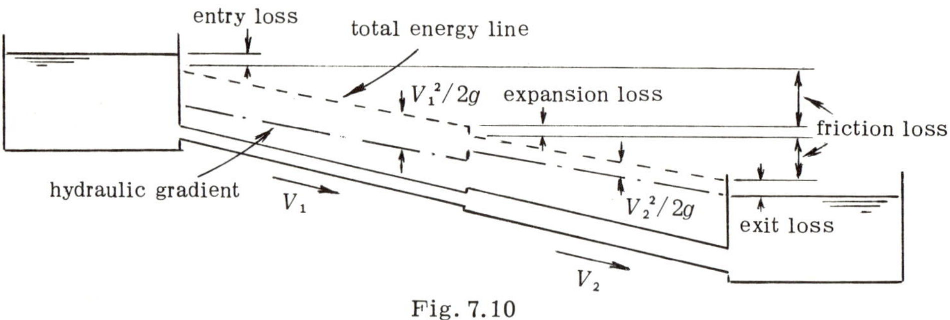

Fig. 7.10

Fig. 7.11 shows a pipe line operating as a siphon over a hill. At A and B, the pressure is atmospheric and between these points, it is below atmospheric. The pressure should never be allowed to fall below about 20 kN/m² or 2 m, otherwise dissolved gases may evolve and flow will cease due to an air-lock.

In most pipe line problems, losses other than pipe friction are negligible.

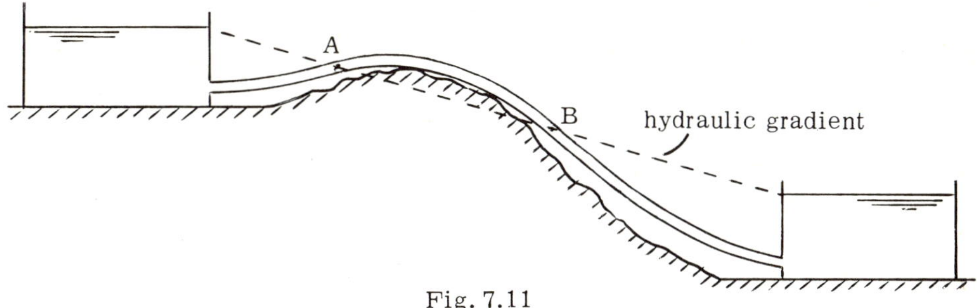

Fig. 7.11

7.13 Combination of pipes These are treated in the same way as combinations of electrical resistances.

(*a*) *Pipes in series* For two pipes in series, Fig. 7.12(*a*), the flow in each pipe is the same,

i.e. $$Q = A_1 V_1 = A_2 V_2$$

and the total loss of head is the sum of the losses in each pipe.

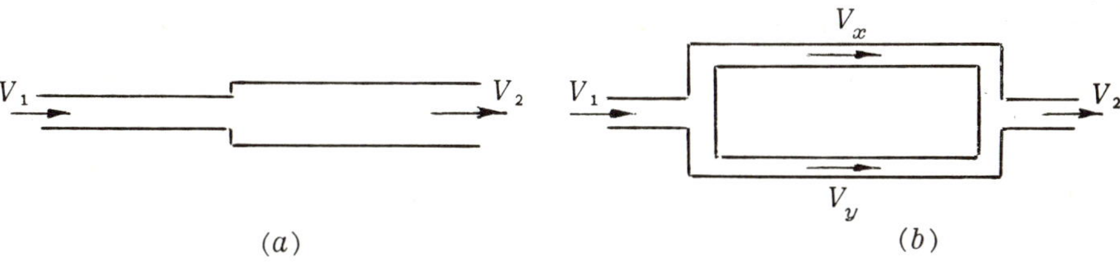

Fig. 7.12

(*b*) *Pipes in parallel* For two pipes in parallel, Fig. 7.12(*b*), the flow in the main pipes is the sum of the flows in the two branches,

i.e. $$Q = A_1 V_1 = A_2 V_2 = A_x V_x + A_y V_y$$

and the loss of head in each branch must be the same.

(*c*) *General case of branched pipes* Fig. 7.13 shows a pipe from a reservoir A which divides into two branches to reservoirs B and C. It is required to find the rate of flow into each of the reservoirs B and C.

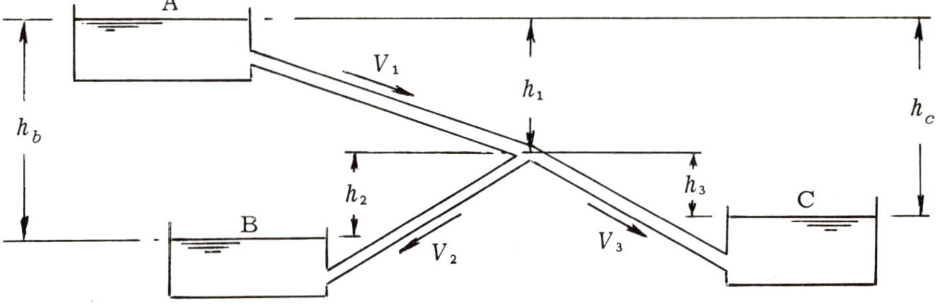

Fig. 7.13

Neglecting all losses other than friction,

$$h_b = h_1 + h_2 = \frac{4f_1 l_1 V_1^2}{2gd_1} + \frac{4f_2 l_2 V_2^2}{2gd_2} \tag{7.11}$$

$$h_c = h_1 + h_3 = \frac{4f_1 l_1 V_1^2}{2gd_1} + \frac{4f_3 l_3 V_3^2}{2gd_3} \tag{7.12}$$

Also $$A_1 V_1 = A_2 V_2 + A_3 V_3 \tag{7.13}$$

The unknown velocities V_1, V_2 and V_3 may be obtained from these equations and hence the discharges calculated.

Such problems may be simplified by writing

$$h_f = \frac{4flV^2}{2gd} = \frac{4flQ^2}{2 \times 9{\cdot}81 \times d \times \left(\frac{\pi}{4}d^2\right)^2} \approx \frac{flQ^2}{3d^5}$$

when equations (7.11), (7.12) and (7.13) become

$$h_b = \frac{f_1 l_1 Q_1^2}{3d_1^5} + \frac{f_2 l_2 Q_2^2}{3d_2^5} \tag{7.14}$$

$$h_c = \frac{f_1 l_1 Q_1^2}{3d_1^5} + \frac{f_3 l_3 Q_3^2}{3d_3^5} \tag{7.15}$$

and $$Q_1 = Q_2 + Q_3 \tag{7.16}$$

The method of solving these equations is shown in Example 6.

7.14 Part-full pipes and open channel flow From the Darcy formula, the hydraulic gradient for a pipe in which the only loss is due to friction is given by

$$i = \frac{h_f}{l} = \frac{4f}{d} \cdot \frac{V^2}{2g} = \frac{f}{\left(\dfrac{\frac{\pi}{4}d^2}{\pi d}\right)} \cdot \frac{V^2}{2g}$$

The term $\dfrac{\frac{\pi}{4}d^2}{\pi d}$ represents the ratio $\dfrac{\text{wetted cross-sectional area}}{\text{wetted perimeter}}$ and is termed the *hydraulic mean depth*. This is denoted by m, so that

$$i = \frac{f}{m} \cdot \frac{V^2}{2g}$$

so that $$V = \sqrt{\frac{2g}{f} \cdot mi} = c\sqrt{mi} \tag{7.17}$$

This is known as the *Chezy formula* and may be used for pipes which are running part-full and for open channels provided that the appropriate value for m is used.

The value of c may be calculated from $\sqrt{\dfrac{2g}{f}}$ but for open channels it is more usually obtained from the Manning formula

$$c = \frac{m^{1/6}}{n}$$

where n, the roughness coefficient, varies from 0·01 for smooth surfaces to 0·03 for rivers.

c has the units $m^{1/2}/s$ so that values for n and c depend upon the system of units used. The values of n quoted above are correct for S.I. units only.

The Chezy formula may be obtained from first principles by considering steady flow in a pipe or channel laid on a gradient i, Fig. 7.14.

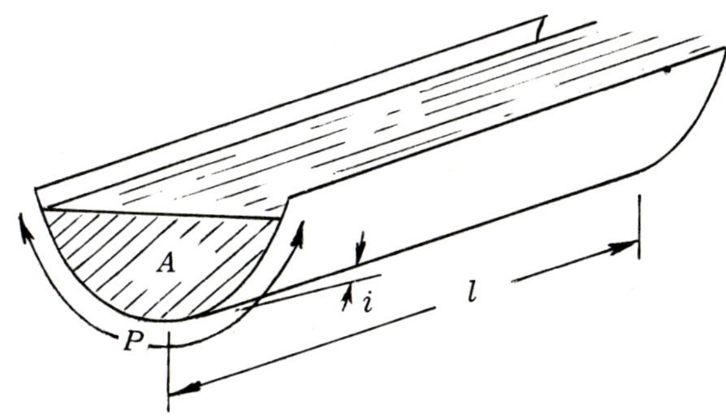

Fig. 7.14

If the length of the channel is l, the wetted cross-section is A and the wetted perimeter is P, the component of the weight of liquid down the plane is equal to the frictional resisting force. For turbulent flow, the frictional resistance may be taken as proportional to the square of the velocity, so that the resistance per unit area, $R = kV^2$

Hence $$\rho g A l \times i = R \times Pl = kV^2 \times Pl$$

from which $$V = \sqrt{\frac{\rho g}{k} \cdot \frac{A}{P} \cdot i}$$

$$= c\sqrt{mi}$$

7.15 Power transmission in pipes with turbulent flow Head represents energy per unit weight, as shown in Art. 4.2. Thus the power P *available* for a flow rate Q under a head h is given by

$$P = \rho g Q h = \rho g A V h$$

For power transmitted a considerable distance by pipeline, the only significant loss of head is that due to friction. Thus, for a friction loss h_f, the *transmitted* power

$$P = \rho g A V (h - h_f)$$

and the transmission efficiency is given by

$$\eta = \frac{\text{power transmitted}}{\text{power available}}$$

$$= \frac{\rho g A V (h - h_f)}{\rho g A V h} = 1 - \frac{h_f}{h}$$

Assuming the friction coefficient f to be constant,

$$h_f = cV^2 \quad \text{where} \quad c = \frac{4fl}{2gd}$$

Thus
$$P = \rho g A V (h - cV^2)$$
$$= \rho g A (hV - cV^3)$$

For the power to be a maximum,

$$\frac{\mathrm{d}}{\mathrm{d}V}(hV - cV^3) = 0$$

i.e.
$$h - 3cV^2 = 0$$

or
$$h_f = \frac{h}{3}$$

For this condition, the transmission efficiency is $66\frac{2}{3}\%$.

Fig. 7.15 shows the variation of P in terms of V. The value of V for maximum power is that which leads to $h_f = h/3$ and hence, for a given pipe length and roughness, the optimum pipe diameter is that which fulfils this relation.

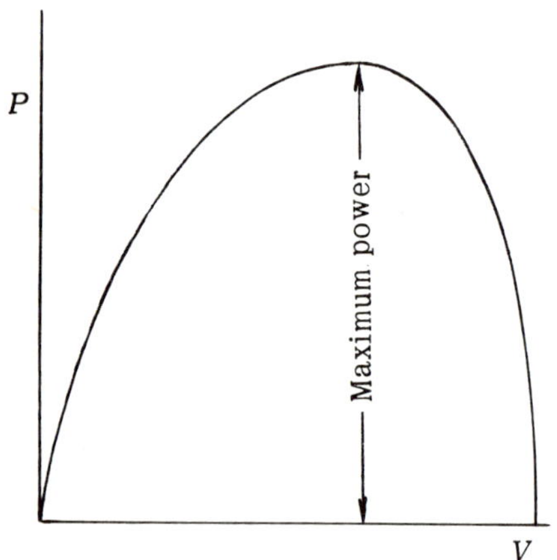

Fig. 7.15

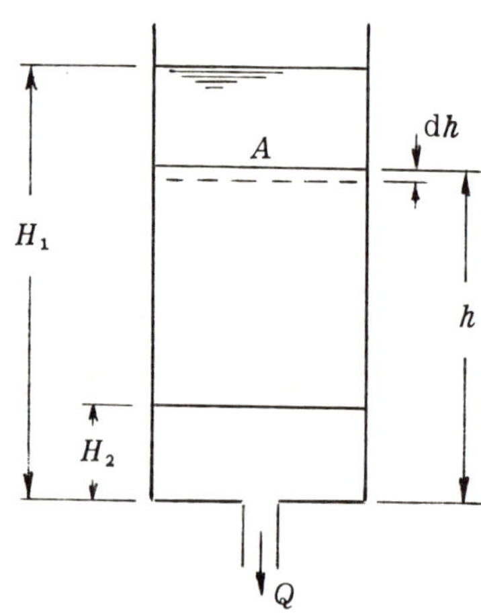

Fig. 7.16

7.16 Quasi-steady flow As stated in Art. 7.2, quasi-steady flow is unsteady but varying so slowly that it can be treated as steady.

Consider the case of a tank of uniform cross-sectional area A, Fig. 7.16, with an orifice of area a in the base.

If the tank is filled to a depth H_1 and then allowed to discharge freely through the orifice, the outflow at any instant

$$Q = C_d a\sqrt{2gh}$$

where h is the instantaneous head over the orifice.

If the head falls a distance $\mathrm{d}h$ in time $\mathrm{d}t$, the reduction in volume of liquid in the tank is equal to the discharge through the orifice.

i.e. $$A\,\mathrm{d}h = -C_d a\sqrt{2gh}\,\mathrm{d}t$$

The negative sign arises because h *decreases* as t *increases*.

Thus $$\mathrm{d}t = \frac{-A}{C_d a\sqrt{2g}}\cdot\frac{\mathrm{d}h}{h^{1/2}}$$

If T is the time taken for the head to fall from H_1 to H_2, then

$$\int_0^T \mathrm{d}t = \frac{-A}{C_d a\sqrt{2g}}\int_{H_1}^{H_2}\frac{\mathrm{d}h}{h^{1/2}}$$

from which $$T = \frac{2A}{C_d a\sqrt{2g}}(\sqrt{H_1} - \sqrt{H_2}) \tag{7.18}$$

If an experiment if performed in which the time T is noted for various values of H_2 as the tank discharges, a graph of T against $\sqrt{H_2}$ should be a straight line of slope $\dfrac{-2A}{C_d a\sqrt{2g}}$, from which C_d may be obtained.

As a further example of quasi-steady flow, consider the flow from reservoir 1 to reservoir 2, Fig. 7.17.

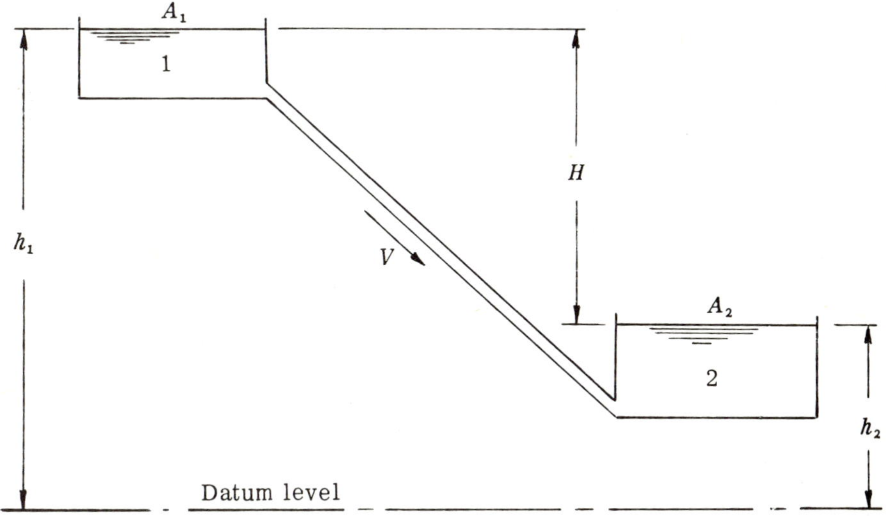

Fig. 7.17

Let the surface areas of the reservoirs be A_1 and A_2 respectively, the cross-sectional area of the pipe be a and the velocity of flow be V. If the level in reservoir 1 falls by $\mathrm{d}h_1$ and that in reservoir 2 rises by $\mathrm{d}h_2$ in time $\mathrm{d}t$, then

$$Q = -A_1\frac{\mathrm{d}h_1}{\mathrm{d}t} = A_2\frac{\mathrm{d}h_2}{\mathrm{d}t} = aV$$

The difference in level, $H = h_1 - h_2$

Therefore, rate of change of level difference,

$$\frac{\mathrm{d}H}{\mathrm{d}t} = \frac{\mathrm{d}h_1}{\mathrm{d}t} - \frac{\mathrm{d}h_2}{\mathrm{d}t}$$

$$= \frac{\mathrm{d}h_1}{\mathrm{d}t}\left(1 + \frac{A_1}{A_2}\right) \quad \text{since } \frac{\mathrm{d}h_2}{\mathrm{d}t} = -\frac{\mathrm{d}h_1}{\mathrm{d}t}\cdot\frac{A_1}{A_2}$$

$$\therefore\ Q = aV = \frac{-A_1}{1+\dfrac{A_1}{A_2}}\cdot\frac{\mathrm{d}H}{\mathrm{d}t}$$

$$= -\frac{A_1A_2}{A_1+A_2}\cdot\frac{\mathrm{d}H}{\mathrm{d}t}$$

The losses of head between the two reservoirs due to friction and other causes are all multiples of $\dfrac{V^2}{2g}$ and so may be represented collectively by $c\dfrac{V^2}{2g}$.

Hence $$H = c\frac{V^2}{2g}$$

or $$V = \sqrt{\frac{2gH}{c}}$$

so that $$a\sqrt{\frac{2gH}{c}} = -\frac{A_1A_2}{A_1+A_2}\cdot\frac{\mathrm{d}H}{\mathrm{d}t}$$

from which $$\mathrm{d}t = -\frac{A_1A_2}{a(A_1+A_2)}\sqrt{\frac{c}{2g}}\frac{\mathrm{d}H}{\sqrt{H}}$$

If T is the time taken for the difference in level to change from H_1 to H_2,

$$\int_0^T \mathrm{d}t = -\frac{A_1A_2}{a(A_1+A_2)}\sqrt{\frac{c}{2g}}\int_{H_1}^{H_2}\frac{\mathrm{d}H}{\sqrt{H}}$$

i.e. $$T = k(\sqrt{H_1} - \sqrt{H_2}) \qquad (7.19)$$

where $$k = \frac{2A_1A_2}{a(A_1+A_2)}\sqrt{\frac{c}{2g}}$$

1. *Oil is pumped through a circular pipe 150 mm diameter and 500 m long and discharges at a level of 20 m above the pump. The oil has a density of 850 kg/m³ and a viscosity of 0·12 N s/m². Determine the power required to pump (a) 25 kg/s of oil, (b) 100 kg/s of oil. Neglect all losses other than pipe friction.*

(*a*)

$$V = \frac{\dot{m}}{\rho A} = \frac{25}{850 \times \frac{\pi}{4} \times 0{\cdot}15^2} = 1{\cdot}662 \text{ m/s}$$

$$\therefore \quad Re = \frac{\rho V d}{\mu} = \frac{850 \times 1{\cdot}662 \times 0{\cdot}15}{0{\cdot}12} = 1\,768$$

Hence the flow is laminar since $Re < 2\,000$.

Thus $$f = \frac{16}{Re} \quad \ldots\ldots \text{ from equation (7.6)}$$

$$= \frac{16}{1\,768} = 0{\cdot}009\,05$$

$$\therefore \quad h_f = \frac{4flV^2}{2gd} \quad \ldots\ldots \text{ from equation (7.1)}$$

$$= \frac{4 \times 0{\cdot}009\,05 \times 500 \times 1{\cdot}662^2}{2 \times 9{\cdot}81 \times 0{\cdot}15} = 16{\cdot}95 \text{ m}$$

Therefore head to be overcome $= 16{\cdot}95 + 20 = 36{\cdot}95$ m

Therefore power required $= \dot{m}gh$

$= 25 \times 9{\cdot}81 \times 36{\cdot}95$

$= 9\,070$ W or $\underline{9{\cdot}07 \text{ kW}}$

(*b*) $$V = 4 \times 1{\cdot}662 = 6{\cdot}65 \text{ m/s}$$

and $$Re = 4 \times 1\,768 = 7\,072$$

Hence the flow is turbulent since $Re > 4\,000$.

Thus $$f = \frac{0{\cdot}079}{Re^{1/4}} \quad \ldots\ldots \text{ from equation (7.7)}$$

$$= \frac{0{\cdot}079}{7\,072^{1/4}} = 0{\cdot}008\,61$$

Thus $$h_f = \frac{4 \times 0{\cdot}008\,61 \times 500 \times 6{\cdot}65^2}{2 \times 9{\cdot}81 \times 0{\cdot}15} = 259 \text{ m}$$

$$\therefore \quad h = 259 + 20 = 279 \text{ m}$$

Therefore power required $= 100 \times 9{\cdot}81 \times 279$

$= 273\,500$ W or $\underline{273{\cdot}5 \text{ kW}}$

2. *At a sudden enlargement in a pipe carrying water, the diameter increases from 320 mm to 640 mm and at the enlargement, the hydraulic gradient rises by 0·12 m. Calculate the mass flow rate of water through the pipe.*

Fig. 7.18 shows the total head and hydraulic gradient just before and after the enlargement. From the figure it will be seen that

$$\frac{V_1^2}{2g} = \frac{V_2^2}{2g} + 0{\cdot}12 + \text{loss of head}$$

From equation (7.8), loss of head in the enlargement $= \dfrac{(V_1 - V_2)^2}{2g}$

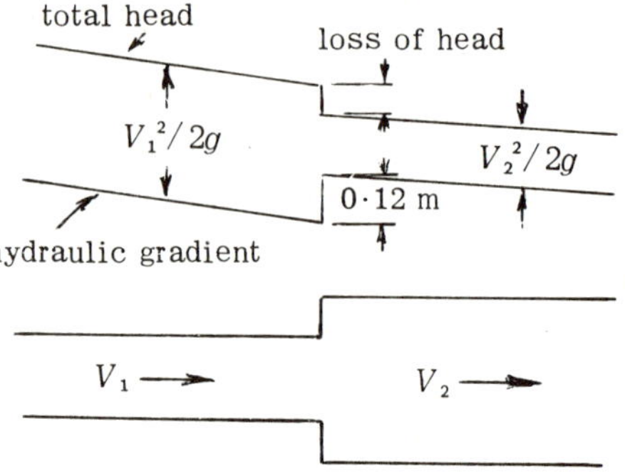

Fig. 7.18

Thus

$$\frac{V_1^2}{2g} = \frac{V_2^2}{2g} + 0{\cdot}12 + \frac{(V_1 - V_2)^2}{2g}$$

But $\qquad A_1 V_1 = A_2 V_2$

so that $\qquad V_2 = V_1\left(\frac{d_1^2}{d_2^2}\right) = \frac{V_1}{4}$

Thus $\qquad V_1^2 = \frac{V_1^2}{16} + 0{\cdot}12 \times 2 \times 9{\cdot}81 + \frac{9}{16}V_1^2$

from which $\qquad V_1 = 2{\cdot}51 \text{ m/s}$

Hence $\qquad \dot{m} = \rho A_1 V_1$

$$= 10^3 \times \frac{\pi}{4} \times 0{\cdot}32^2 \times 2{\cdot}51 = \underline{202 \text{ kg/s}}$$

3. *Two reservoirs, Fig. 7.19, whose difference in level is 15 m are connected by a pipe PQR, 1500 m in length, which has its highest point Q 1·7 m below the level in the upper reservoir. The pipe is 200 mm in diameter and $f = 0{\cdot}005$.*

Find the length PQ so that the pressure at Q is 3 m below atmospheric.

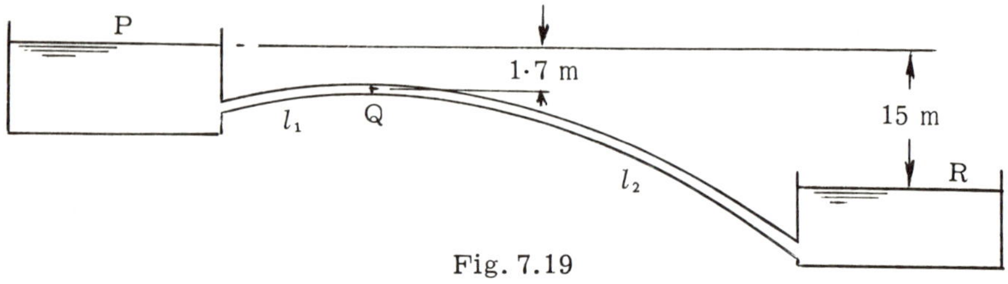

Fig. 7.19

The losses of head at entry and exit are $0{\cdot}5\dfrac{V^2}{2g}$ and $\dfrac{V^2}{2g}$ respectively.

Applying Bernoulli's equation to points P and Q, taking the level at R as datum and the pressure of the atmosphere as p,

$$\frac{p}{\rho g} + 15 = \left(\frac{p}{\rho g} - 3\right) + \frac{V^2}{2g} + 13{\cdot}3 + \frac{4fl_1V^2}{2gd} + 0{\cdot}5\frac{V^2}{2g}$$

from which $$4{\cdot}7 = \frac{V^2}{2g}\left(1{\cdot}5 + \frac{4 \times 0{\cdot}005 l_1}{0{\cdot}2}\right)$$

$$= \frac{V^2}{2g}\left(1{\cdot}5 + \frac{l_1}{10}\right) \qquad (1)$$

Applying Bernoulli's equation to points P and R,

$$\frac{p}{\rho g} + 15 = \frac{p}{\rho g} + \frac{4f(l_1 + l_2)V^2}{2gd} + 0{\cdot}5\frac{V^2}{2g} + \frac{V^2}{2g}$$

from which $$15 = \frac{V^2}{2g}\left(1{\cdot}5 + \frac{4 \times 0{\cdot}005 \times 1\,500}{0{\cdot}2}\right)$$

$$= 151{\cdot}5\frac{V^2}{2g} \qquad (2)$$

Thus, from equations (1) and (2),

$$\frac{15}{4{\cdot}7} = \frac{151{\cdot}5}{1{\cdot}5 + \dfrac{l_1}{10}}$$

from which $$l_1 = \underline{460 \text{ m}}$$

4. *A pipe line comprising two pipes of the same length and diameter delivers water from a reservoir to a turbine. Each pipe is 4 000 m in length and $f = 0{\cdot}006$. Calculate the diameter of the pipes if the turbine is to develop 4 MW with a gross head of 170 m. All losses except pipe friction may be ignored.*

Each pipe delivers half the water to produce 2 MW and from Art. 7.15, the maximum power will be achieved when the head lost is one-third of the available head.

i.e. $$h_f = \frac{170}{3} = \frac{4flV^2}{2gd} = \frac{4 \times 0{\cdot}006 \times 4\,000V^2}{2gd}$$

from which $$\frac{V^2}{d} = 11{\cdot}6 \qquad (1)$$

Power available $$= \rho gAV(h - h_f)$$

i.e. $$2 \times 10^6 = 10^3 \times 9{\cdot}81 \times \frac{\pi}{4}d^2V\left(170 - \frac{170}{3}\right)$$

from which $$Vd^2 = 2{\cdot}29 \qquad (2)$$

Thus, from equations (1) and (2),

$$d^5 = \frac{2{\cdot}29^2}{11{\cdot}6} = 0{\cdot}452$$

$$\therefore \; d = \underline{0{\cdot}854 \text{ m}}$$

5. *A pipe line 2 000 m in length has two pipes in parallel. One pipe is 0·4 m diameter and the other is 0·5 m diameter. The total flow rate is 0·8 m³/s and for both pipes, f = 0·006. Determine the loss of head in the system and the flow rate in each pipe.*

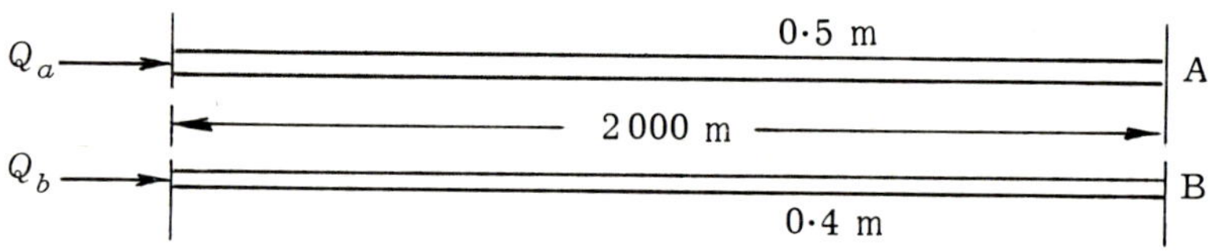

Fig. 7.20

For pipes in parallel, the loss of head in each must be the same. Thus, referring to Fig. 7.20,

$$h_f = \frac{4flV_a^2}{2gd_a} = \frac{4flV_b^2}{2gd_b}.$$

$$\therefore \; V_a = V_b\sqrt{\frac{d_a}{d_b}} = V_b\sqrt{\frac{0{\cdot}5}{0{\cdot}4}} = 1{\cdot}12\,V_b \qquad (1)$$

$$Q_a + Q_b = Q$$

i.e. $\frac{\pi}{4}\times 0{\cdot}5^2\,V_a + \frac{\pi}{4}\times 0{\cdot}4^2\,V_b = 0{\cdot}8$

i.e. $1{\cdot}5625\,V_a + V_b = 6{\cdot}36 \qquad (2)$

From equations (1) and (2),

$$V_a = 2{\cdot}59 \text{ m/s} \quad \text{and} \quad V_b = 2{\cdot}32 \text{ m/s}$$

$$\therefore \; h_f = \frac{4flV_a^2}{2gd_a}$$

$$= \frac{4\times 0{\cdot}006\times 2\,000\times 2{\cdot}59^2}{2\times 9{\cdot}81\times 0{\cdot}5} = \underline{32{\cdot}82 \text{ m}}$$

$$Q_a = \frac{\pi}{4}\times 0{\cdot}5^2\times 2{\cdot}59 = \underline{0{\cdot}51 \text{ m}^3/\text{s}}$$

and $$Q_b = \frac{\pi}{4}\times 0{\cdot}4^2\times 2{\cdot}32 = \underline{0{\cdot}29 \text{ m}^3/\text{s}}$$

6. *Water flows from a reservoir through a pipe 150 mm diameter and 180 m long to a point below the surface of the reservoir where it branches into two pipes, each 100 mm diameter. One of the pipes is 48 m long discharging to atmosphere at a point 18 m below reservoir level and the other 60 m long discharging to atmosphere 24 m below reservoir level.*

Assuming that $f = 0{\cdot}008$, calculate the discharge from each pipe, neglecting all losses other than friction.

The system is shown in Fig. 7.21.

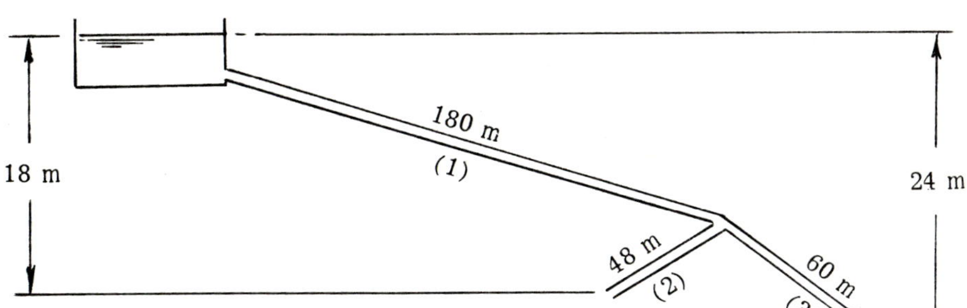

Fig. 7.21

From equations (7.14), (7.15) and (7.16),

$$18 = \frac{0{\cdot}008 \times 180\,Q_1^2}{3 \times 0{\cdot}15^5} + \frac{0{\cdot}008 \times 48\,Q_2^2}{3 \times 0{\cdot}10^5}$$

$$= 6\,320\,Q_1^2 + 12\,800\,Q_2^2 \qquad (1)$$

$$24 = \frac{0{\cdot}008 \times 180\,Q_1^2}{3 \times 0{\cdot}15^5} + \frac{0{\cdot}008 \times 60\,Q_3^2}{3 \times 0{\cdot}10^5}$$

$$= 6\,320\,Q_1^2 + 16\,000\,Q_3^2 \qquad (2)$$

and $$Q_1 = Q_2 + Q_3 \qquad (3)$$

From equation (1), $0{\cdot}002\,85 = Q_1^2 + 2{\cdot}025\,Q_2^2$

or $$\frac{0{\cdot}002\,85}{Q_1^2} = 1 + 2{\cdot}025\,n^2 \qquad \text{where } n = \frac{Q_2}{Q_1}$$

From equation (2), $0{\cdot}003\,8 = Q_1^2 + 2{\cdot}53\,Q_3^2$

or $$\frac{0{\cdot}003\,8}{Q_1^2} = 1 + 2{\cdot}53\left(\frac{Q_3}{Q_1}\right)^2$$

$$= 1 + 2{\cdot}53\left(1 - \frac{Q_2}{Q_1}\right)^2 \quad \text{from equation (3)}$$

$$= 1 + 2{\cdot}53\,(1 - n)^2$$

$$\text{Thus} \qquad \frac{0{\cdot}002\,85}{0{\cdot}003\,8} = \frac{1 + 2{\cdot}025\,n^2}{1 + 2{\cdot}53\,(1-n)^2}$$

$$\text{from which} \quad n^2 + 30{\cdot}7n - 13{\cdot}35 = 0$$

$$\text{Thus} \qquad n = 0{\cdot}425$$

$$\therefore \quad \frac{0{\cdot}002\,85}{Q_1^2} = 1 + 2{\cdot}025 \times 0{\cdot}425^2 = 1{\cdot}365$$

$$\therefore \quad Q_1 = \underline{0{\cdot}045\,7 \text{ m}^3/\text{s}}$$

$$Q_2 = 0{\cdot}425 \times 0{\cdot}045\,7$$

$$= \underline{0{\cdot}019\,43 \text{ m}^3/\text{s}}$$

7. *Fig. 7.22 shows the cross-section of an open channel, the lower part being a semicircle. The channel is laid on a slope of 1 in 2 500 and the Chezy constant c = 52. Calculate the volume flow rate when the maximum depth is 0·6 m and also the depth for this rate to be doubled.*

When the maximum depth is 0·6 m,

$$x = 0{\cdot}1 \text{ m}$$

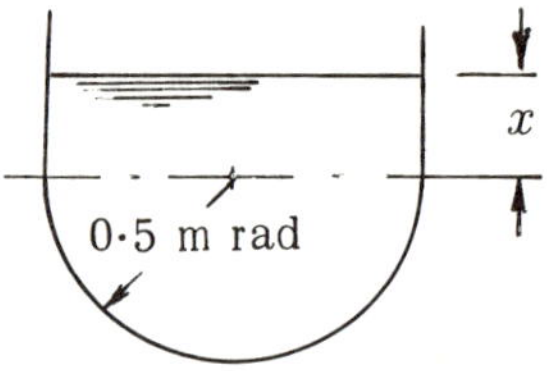

Therefore wetted area,
$$A = \frac{1}{2} \times \frac{\pi}{4} \times 1^2 + 1 \times 0{\cdot}1$$

$$= 0{\cdot}493 \text{ m}^2$$

Wetted perimeter,
$$P = \frac{\pi}{2} \times 1 + 2 \times 0{\cdot}1$$

$$= 1{\cdot}77 \text{ m}$$

Hydraulic mean depth,
$$m = \frac{A}{P} = \frac{0{\cdot}493}{1{\cdot}77} = 0{\cdot}279$$

$$V = c\sqrt{mi} \quad . \quad . \quad . \quad . \quad \text{from equation (7.17)}$$

$$= 52\sqrt{0{\cdot}279 \times \frac{1}{2\,500}}$$

$$= 0{\cdot}548 \text{ m/s}$$

$$\therefore \quad Q = AV = 0{\cdot}493 \times 0{\cdot}548 = \underline{0{\cdot}271 \text{ m}^3/\text{s}}$$

Fig. 7.22

In terms of x,
$$A = \frac{1}{2} \times \frac{\pi}{4} \times 1^2 + 1 \times x = 0{\cdot}393 + x$$

$$P = \frac{\pi}{2} \times 1 + 2 \times x = 1{\cdot}57 + 2x$$

$$\therefore \quad m = \frac{0{\cdot}393 + x}{1{\cdot}57 + 2x}$$

When the flow rate is doubled,

$$V = \frac{Q}{A} = \frac{2 \times 0{\cdot}271}{0{\cdot}393 + x}$$

$$\therefore \quad \frac{0{\cdot}542}{0{\cdot}393 + x} = 52\sqrt{\frac{0{\cdot}393 + x}{1{\cdot}57 + 2x} \times \frac{1}{2\,500}}$$

from which $x = 0{\cdot}494$ m

Thus new maximum depth $= 0{\cdot}5 + 0{\cdot}494 = \underline{0{\cdot}994 \text{ m}}$

8. *A rectangular tank 6 m by 6 m contains water to a depth of 5 m. The tank is connected by a horizontal pipe 300 m long and 75 mm diameter to a second rectangular tank 5 m by 5 m, containing water to a depth of 2 m. The friction coefficient for the pipe is 0·01. How long does it take to lower the depth in the first tank to 4 m?*

The arrangement of the tanks is shown in Fig. 7.23.

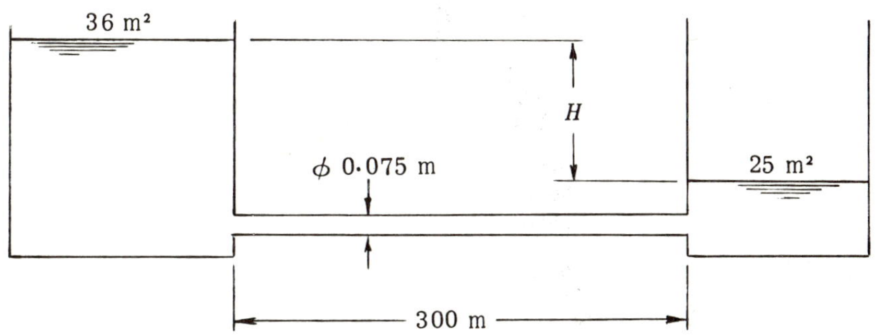

Fig. 7.23

From equation (7.19), $T = k(\sqrt{H_1} - \sqrt{H_2})$

where $k = \dfrac{2A_1A_2}{a(A_1 + A_2)}\sqrt{\dfrac{c}{2g}}$

In this example, the losses are due to pipe friction and losses at entry and exit, so that

$$H = \left(\frac{4fl}{d} + 0{\cdot}5 + 1\right)\frac{V^2}{2g}$$

$$= \left(\frac{4 \times 0{\cdot}01 \times 300}{0{\cdot}075} + 1{\cdot}5\right)\frac{V^2}{2g}$$

$$= 161{\cdot}5\,\frac{V^2}{2g}$$

i.e. $$c = 161{\cdot}5$$

Initial head difference, $H_1 = 5 - 2 = 3$ m

When level in tank 1 falls by 1 m, level in tank 2 rises by $1 \times \frac{36}{25} = 1{\cdot}44$ m.

$\therefore$ final head difference, $H_2 = 4 - 3{\cdot}44 = 0{\cdot}56$ m

Hence
$$T = \frac{2 \times 36 \times 25}{\frac{\pi}{4} \times 0{\cdot}075^2\,(36+25)}\sqrt{\frac{161{\cdot}5}{2 \times 9{\cdot}81}}(\sqrt{3} - \sqrt{0{\cdot}56})$$

$$= 18\,850 \text{ s} \quad \text{or} \quad \underline{5{\cdot}236 \text{ h}}$$

If losses other than friction are ignored, $c = 160$ and $T = 18\,760$, a difference of less than 0·5%. In most cases where long pipes are involved, only friction losses are relevant.

9. Oil of density 850 kg/m³ and viscosity 0·13 kg/m s flows in a pipe of diameter 25 mm at a velocity of 10 m/s. Determine whether flow is laminar or turbulent.

If the critical Reynolds number is taken to be 2000, determine the mass flow rate through the pipe when flow ceases to be laminar.

(*Ans.*: Laminar $[Re = 1\,634]$; 5·1 kg/s)

10. Water flows in a 6 mm diameter pipe at a mean velocity of 0·2 m/s; the viscosity of the water is 0·001 3 N s/m². Determine the loss of head in a 50 m length of pipe and the power required to overcome the loss. (*Ans.*: 1·18 m; 0·065 5 W)

11. A laminar flow of liquid, of viscosity μ, passes through a circular pipe of diameter d with velocity V. Show that the loss of pressure per unit length of pipe is $32V\mu/d^2$. It may be assumed that in laminar flow, $f = 16/Re$ where f is a friction constant defined by f = viscous shear stress/$\frac{1}{2}\rho V^2$.

Calculate the power required to force oil of viscosity 0·25 N s/m² through a 100 mm diameter pipe of length 100 m at a velocity of 0·6 m/s. (*Ans.*: 226 W)

12. A fluid of density 800 kg/m³ and viscosity 0·008 N s/m² flows along a smooth pipe, 8 mm diameter. The pipe is 4 m long and the mean velocity of the fluid is 3·5 m/s. Determine the friction factor and the loss of head in the pipe.

If the same flow rate of fluid is to pass along a smooth pipe of diameter (*a*) 4 mm, (*b*) 16 mm, determine the new loss of head in each case.

(*Ans.*: 0·010 9; 22·9 m; 365 m; 0·444 m)

13. Water at 20°C flows in a smooth pipe 8 m long and 50 mm diameter. The flow rate is 0·02 m³/s. Determine the pressure change in the pipe when the pipe is inclined at 30° to the horizontal and the flow is (*a*) upwards, (*b*) downwards.

(*Ans.*: 138·2 kN/m²; 59·8 kN/m²)

14. A fluid is discharged from a tank through a pipe 700 m long. The first 200 m of the pipe is 80 mm in diameter and the remainder is 160 mm in diameter. The pipe outlet is 40 m below the inlet. The friction coefficient f for both portions of the pipe is 0·003 8, the loss of head at entry to the pipe is $0{\cdot}5\,V^2/2g$ and the loss of head at exit is $V^2/2g$ where V is the relevant pipe velocity. Calculate the flow rate through the pipe, ignoring the loss at enlargement. (*Ans.*: 0·021 7 m³/s)

15. A vertical pipe AB conveys water. The diameter at A is 120 mm and at B, 3 m below A, it is 60 mm. The loss of head in a valve between A and B is of the form $kV^2/2g$, where V is the velocity at A. When the flow rate is 0·03 m³/s upwards, the pressure at B is 12 kN/m² greater than that at A. Find the value of k.

(*Ans.*: 10·1)

16. A sudden enlargement occurs in a circular horizontal pipe which conveys 0·11 m^3/s of oil of density 850 kg/m^3. The enlargement is from 150 mm diameter to 210 mm and the pressure just before the enlargement is 0·1 MN/m^2. Calculate the power required to overcome the loss in the enlargement and the pressure after the enlargement. (*Ans.*: 0·434 kW; 96·05 kN/m^2)

17. Water is siphoned from a tank by a bent pipe, 30 m long and 30 mm diameter. The highest point in the pipe is 2 m below the water level in the tank and the pipe discharges to atmosphere 15 m below the water level. If the highest point in the pipe is one-third of the way along it, determine the pressure at this point.

$f = 0{\cdot}004\,8$ and entry and exit losses may be ignored.

(*Ans.*: 34 kN/m^2 below atmospheric)

18. A reservoir discharges into a second reservoir through a pipe 0·4 m diameter and 2 100 m long. The reservoirs are of equal depth and the difference in surface levels is 9 m. If the level of the pipe midway between the reservoirs is 17 m below the water level in the highest reservoir, calculate the gauge pressure at this point and the flow rate through the pipe. $f = 0{\cdot}005$. (*Ans.*: 353 kW; 0·343 m)

19. A horizontal pipe 30 m long conveys 1·2 m^3/s of water for power. At entry to the pipe, a pump gives a head of 45 m. Determine the maximum power available at exit and the diameter of pipe required, ignoring all losses except friction. $f = 0{\cdot}005$. (*Ans.*: 353 kW; 0·343 m)

20. A 1 m diameter pipe 4 000 m long connects two reservoirs and its slope is 1 in 1 500. When the level of the reservoirs is low, the pipe runs part full and with a depth of 0·5 m, the flow rate is 0·3 m^3/s. Determine the value of f for the pipe and the flow rate when the pipe is full.

Consider also what happens when the pipe runs (*a*) one-quarter full, and (*b*) (*b*) three-quarters full. (*Ans.*: 0·005 6; 0·6 m^3/s)

21. Two pipes, one 50 mm diameter and one 100 mm diameter, each 30 m long, are connected in parallel between two reservoirs whose difference in level is 10 m. Find the flow rate through each pipe if $f = 0{\cdot}008$ for both and entry and exit losses are to be included. (*Ans.*: 0·006 m^3/s; 0·033 m^3/s)

22. Two reservoirs, having a constant difference in water level of 65 m, are connected by a straight pipe 0·2 m diameter and 4 km long. The pipe is tapped at a point distant 1·5 km from the upper reservoir and water is drawn off at the rate of 0·04 m^3/s.

If the friction coefficient f is 0·009, determine the rate at which water enters the lower reservoir, neglecting all losses except pipe friction.

Sketch the hydraulic gradient for the pipe. (*Ans.*: 0·021 8 m^3/s)

23. A pipe line, 0·8 m diameter and 4 km long, delivers 0·8 m^3/s of water at 20°C. If the last 2 km of the line are replaced by two pipes in parallel, each 0·5 m diameter, what will be the percentage change in loss of head? For all pipes f may be assumed to be twice the value for a smooth pipe. Losses other than that due to friction may be ignored. (*Ans.*: 92% increase)

24. A reservoir A supplies two reservoirs B and C, the water surfaces of B and C being 20 m and 30 m respectively below that of A. A pipe 150 mm diameter is used for the first 270 m from A and this then branches into two pipes, each 100 mm diameter; the pipe to reservoir B is 65 m long while that to reservoir C is 100 m long.

Calculate the discharge to reservoirs B and C, taking $f = 0{\cdot}007\,5$ for all pipes. Neglect all losses except those due to friction. (*Ans.*: 0·007 m^3/s; 0·028 m^3/s)

25. A cylindrical water tank 0·8 m in diameter discharges through a sharp-edged circular orifice, 15 mm diameter, in the base of the tank. If the initial water level is 0·75 m above the base and the coefficient of discharge of the orifice is 0·62, calculate

(*a*) the time taken to empty the tank,
(*b*) the water level after 500 s. (*Ans.*: 1 792 s; 0·39 m)

26. A vertical cylindrical tank, 0·3 m diameter, has a 19 mm diameter orifice in the base. The tank is filled with water to a depth of 1 m and allowed to discharge through the orifice. The table shows readings of time t at head h above the orifice.

h (mm)	1 000	920	835	750	670	580	500	420	335
t (s)	0	7·6	15·5	23·8	32·8	42·1	52·2	63·3	75·3

Determine the coefficient of discharge of the orifice. (*Ans*.: 0·632)

27. The petrol tank drain plug falls out of the tank in a car, leaving a circular orifice 25 mm diameter in the tank base. The tank is rectangular in cross-section, 1 m × 0·2 m.

The motorist had just filled his tank with 66 litres of petrol. Ignoring the small amount of fuel consumed in motoring, determine how far the car will travel at 45 km/h before it runs out of petrol. The coefficient of discharge of the orifice is 0·6. (*Ans*.: 2·2 km)

28. A cylindrical tank, 3 m diameter, is emptied through an open ended pipe fitted in the bottom of the tank. The pipe is 50 mm diameter and 30 m long, with its open end 10 m below the bottom of the tank. Taking the friction coefficient in the pipe as 0·007 and allowing for losses at entry and exit, find the time taken to lower the level in the tank from 4 m to 1 m. (*Ans*.: 49·26 min)

29. A tank of constant cross-sectional area 8 m^2 is connected to another tank of constant cross-sectional area 4 m^2 by a pipe 50 mm diameter and 130 m long. The friction coefficient for the pipe is 0·01. Neglecting losses other than that due to friction, find the time taken for 2·25 m^3 of water to flow from the large tank to the small tank if the initial level in the large tank is 2 m above that in the small tank. (*Ans*.: 35·39 min)

8 Dimensional analysis

8.1 Introduction The fundamental dimensions of engineering quantities are mass (M), length (L), time (T) and temperature (θ) and all equations must have the same dimensions on either side. This provides a method of determining the nature of a relationship which may be too complex for simple analysis. This *dimensional analysis* method will give only part of the solution and experimental work is necessary to complete it.

Although dimensional analysis may be used in the solution of problems in many fields, it is restricted here to fluid mechanics, in which temperature is not involved and the fundamental dimensions are then reduced to three: M, L and T. The following table shows the dimensions of quantities which commonly arise in fluid mechanics:

Quantity	Dimensions	Quantity	Dimensions
Area	L^2	Pressure, stress	M/LT^2
Volume	L^3	Volume flow rate	L^3/T
Velocity	L/T	Work, energy	ML^2/T^2
Acceleration	L/T^2	Power	ML^2/T^3
Momentum	ML/T	Dynamic viscosity	M/LT
Force	ML/T^2	Kinematic viscosity	L^2/T
Angular velocity	$1/T$	Surface tension	M/T^2
Mass flow rate	M/T	Density	M/L^3

To illustrate the principle, consider the dimensions of the continuity equation

$$\dot{m} = \rho A V$$

Using the data in the table, the dimensions on the two sides are

$$\left[\frac{M}{T}\right] = \left[\frac{M}{L^3}\right][L^2]\left[\frac{L}{T}\right] = \left[\frac{M}{T}\right]$$

Thus it can be seen that the dimensions on either side are the same.

8.2 Rayleigh's method of dimensional analysis The variables considered to be involved in a problem are chosen and the equation is expressed in the form

$$X = \phi(X_1, X_2, X_3, \ldots)$$

$$= k(X_1^a, X_2^b, X_3^c, \ldots)$$

where X_1, X_2, X_3, etc., are the variables, ϕ represents 'some function of', k is a constant and a, b, c, d, etc., are unknown indices.

The dimensions of each of the variables are put into the equation and the corresponding dimensions in the two sides are equated, i.e.,

dimension of X = [dimension of $X_1]^a$ [dimension of $X_2]^b$ [dimension of $X_3]^c \ldots$

If all four fundamental dimensions are involved, four indices may be eliminated. Where there are not more than four variables, the complete form of the equation may be determined, except for numerical constants but when there are more than four variables, the resulting equation can only be expressed in terms of dimensionless groups and different, but equally valid, forms may be obtained by eliminating different indices.

Certain dimensionless groups occur frequently in hydraulics problems and are given special names such as Reynolds number and Froude number.

This method is illustrated in Examples 1 and 2 and in Arts. 9.4 and 9.5.

8.3 Buckingham's Π-theorem The principle of dimensional analysis is also stated in Buckingham's Π-theorem * as follows:

If a physical problem involves n variables, $X_1, X_2, \ldots, X_n$, then

$$\phi(X_1, X_2, \ldots, X_n) = 0 \tag{8.1}$$

If p is the number of fundamental dimensions, such as M, L and T, in the problem, the n variables may also be arranged in $(n-p)$ independent dimensionless groups, each denoted by Π.

Then $$\phi(\Pi_1, \Pi_2, \ldots, \Pi_{n-p}) = 0 \tag{8.2}$$

Each Π term involves some of the X terms and ϕ represents 'some function of'.

The procedure is as follows:

(*a*) Select any p variables from the n available variables, which *must* include between them all the p fundamental dimensions.

(*b*) Select one other variable and form a dimensionless group from these $p+1$ variables.

(*c*) Repeat (*b*) for each other variable in turn.

Thus, for example, select X_1, X_2 and X_3, containing between them M, L and T and combine with X_4 to form Π_1. Then combine X_1, X_2 and X_3 with X_5 to form Π_2 and so on up to X_n to form Π_{n-p}.

$$\begin{aligned} \Pi_1 &= \phi_1(X_1, X_2, X_3, X_4) \\ \Pi_2 &= \phi_2(X_1, X_2, X_3, X_5) \\ &\;\;\vdots \\ \Pi_{n-p} &= \phi_{n-p}(X_1, X_2, X_3, X_n) \end{aligned} \tag{8.3}$$

* For proof, see '*E. Buckingham* — Model Experiments and the Form of Empirical Equations', Trans. ASME, Vol. 37, 1915.

Equations (8.3) are solved individually as in Art. 8.2, with the index for the variables $X_4, X_5, \ldots, X_n$ taken as unity in each case.

This method is illustrated in Examples 1 and 2. It can be seen that, although longer than Rayleigh's method, it is simpler for problems involving many variables since each group is produced individually.

8.4 Arrangement of dimensionless groups By Rayleigh's method, it is possible to produce different dimensionless groups by varying the indices chosen to be eliminated. Similarly, by Buckingham's method, different dimensionless groups may be produced by varying the three chosen variables from the n available; this is of no significance since, because the groups are dimensionless, they may be multiplied or divided amongst themselves until the required arrangement is obtained. Any new groups formed in this way will still be dimensionless (see Examples 1 and 2).

In the case of Rayleigh's method, in which a given relationship such as

$$R = \rho V^2 D^2 \phi(Re\,,\, Fr) \;\; *$$

is to be derived from the variables D, V, ρ, μ and g, the result may be obtained directly by eliminating the indices of the terms which appear outside the brackets, i.e. the indices of ρ, V and D. It may not always be possible to achieve this aim in multi-variable problems, so that some re-arrangement will still be needed but this choice minimizes the work.

A problem sometimes occurs with the Buckingham method when selecting the three initial variables. If it is possible to form a dimensionless group from the three chosen variables, the fourth variable which is added is not involved. This has an index of 1, which will automatically produce meaningless equations and hence it is necessary to change one of the three chosen variables to obtain a solution (see Example 1).

1. *A circular shaft of diameter D rotates at a speed N and is supported by a journal bearing with diametral clearance C. The viscosity of the lubricant is μ and the bearing pressure is p. Obtain an expression for the friction coefficient f, defined by the ratio friction force/bearing load, in terms of μ, N, p, D and C.*

(*a*) **Rayleigh's method**

$$f = \phi(\mu, N, p, D, C) = k(\mu^a N^b p^c D^d C^e)$$

Substituting dimensions,

$$0 = \left[\frac{M}{LT}\right]^a \left[\frac{1}{T}\right]^b \left[\frac{M}{LT^2}\right]^c [L]^d\,[L]^e$$

Equating powers of M: $\quad 0 = a + c \qquad (1)$

Equating powers of L: $\quad 0 = -a - c + d + e \qquad (2)$

Equating powers of T: $\quad 0 = -a - b - 2c \qquad (3)$

* See Art. 9.5.

Three indices may be eliminated from equations (1), (2) and (3). One solution is $a = b$, $c = -b$ and $e = -d$, so that

$$f = k(\mu^b N^b p^{-b} D^d C^{-d})$$

$$= k\left[\frac{D}{C}\right]^d \left[\frac{\mu N}{p}\right]^b$$

or

$$f = \phi\left[\frac{D}{C}, \frac{\mu N}{p}\right]$$

(*b*) **Buckingham's method**

$$\phi(f, \mu, N, p, D, C) = 0$$

Three dimensionless groups will be formed from the six variables but one of these is the friction coefficient, f, leaving two further groups Π_1 and Π_2.

Thus $\phi(\Pi_1, \Pi_2, f) = 0$

or $f = \phi(\Pi_1, \Pi_2)$

Π_1 and Π_2 will be obtained from $\phi(\mu, N, p, D, C) = 0$.

Choose μ, N and D to involve the dimensions M, L and T.

Then $\Pi_1 = \phi_1(\mu, N, D, p) = \mu^{a_1} N^{b_1} D^{c_1} p$

Since each Π group is dimensionless, then, substituting dimensions:

$$0 = \left[\frac{M}{LT}\right]^{a_1} \left[\frac{1}{T}\right]^{b_1} [L]^{c_1} \left[\frac{M}{LT^2}\right]$$

Equating powers of M: $0 = a_1 + 1$

Equating powers of L: $0 = -a_1 + c_1 - 1$

Equating powers of T: $0 = -a_1 - b_1 - 2$

Hence $a_1 = -1$, $b_1 = -1$ and $c_1 = 0$

so that $\Pi_1 = \mu^{-1} N^{-1} p = \dfrac{p}{\mu N}$

Similarly $\Pi_2 = \phi_2(\mu, N, D, C) = \mu^{a_2} N^{b_2} D^{c_2} C$

$$\therefore\ 0 = \left[\frac{M}{LT}\right]^{a_2} \left[\frac{1}{T}\right]^{b_2} [L]^{c_2} [L]$$

Equating powers of M: $0 = a_2$

Equating powers of L: $0 = -a_2 + c_2 + 1$

Equating powers of T: $0 = -a_2 - b_2$

Hence $a_2 = 0$, $b_2 = 0$ and $c_2 = -1$

so that $$\Pi_2 = D^{-1}C = \frac{C}{D}$$

Thus $$f = \phi(\Pi_1, \Pi_2) = \phi\left[\frac{p}{\mu N}; \frac{C}{D}\right]$$

It will be seen that, in this case, the groups are inverted compared with those obtained by Rayleigh's method. However, this is of no significance since, because they are dimensionless, they may be inverted without loss of validity.

This example may be used to demonstrate the problem that arises if the three chosen variables in the Buckingham method form a dimensionless group of their own. It can be seen from the results that μ, N and p could have been selected since they involve M, L and T but they form their own group.

Combining these variables with D, then

$$\Pi_1 = \phi_1(\mu, N, p, D) = \mu^{a_1} N^{b_1} p^{c_1} D$$

Substituting dimensions,

$$0 = \left[\frac{M}{LT}\right]^{a_1}\left[\frac{1}{T}\right]^{b_1}\left[\frac{M}{LT^2}\right]^{c_1}[L]$$

Equating powers of M, L and T respectively gives

$$0 = a_1 + c_1$$

$$0 = -a_1 - c_1 + 1$$

and $$0 = -a_1 - b_1 - 2c_1$$

The first two equations lead to the statement that $1 = 0$.

The results from this combination are therefore meaningless and it is necessary to change one of the three chosen variables.

2. *An orifice of diameter d is used to measure the rate of flow Q of a fluid of viscosity μ and density ρ along a pipe of diameter D. The pressure drop across the orifice is p. Show that*

$$Q = d^2\sqrt{\frac{p}{\rho}}\,\phi\left[\frac{\mu}{d\sqrt{\rho p}}, \frac{D}{d}\right]$$

(*a*) **Rayleigh's method**

$$Q = \phi(d, \rho, \mu, D, p) = k(d^a \rho^b \mu^c D^d p^e)$$

Substituting dimensions,

$$\frac{L^3}{T} = [L]^a\left[\frac{M}{L^3}\right]^b\left[\frac{M}{LT}\right]^c[L]^d\left[\frac{M}{LT^2}\right]^e$$

Equating powers of M: $0 = b + c + e$

Equating powers of L: $3 = a - 3b - c + d - e$

Equating powers of T: $-1 = -c - 2e$

Since the required expression has the variables d, p and ρ as multipliers, it is necessary to eliminate the indices of these terms, i.e. a, e and b.

Thus $$a = 2 - c - d$$

$$e = \frac{1}{2} - \frac{c}{2}$$

and $$b = -\frac{1}{2} - \frac{c}{2}$$

Hence $$Q = k\,\{d^{2-c-d}\rho^{-\frac{1}{2}-\frac{c}{2}}\mu^{c}D^{d}p^{\frac{1}{2}-\frac{c}{2}}\}$$

$$= kd^2\sqrt{\frac{p}{\rho}}\left[\frac{\mu}{d\sqrt{\rho p}}\right]^{c}\left[\frac{D}{d}\right]^{d}$$

$$= d^2\sqrt{\frac{p}{\rho}}\,\phi\left[\frac{\mu}{d\sqrt{\rho p}}, \frac{D}{d}\right]$$

(*b*) **Buckingham's method**

$$\phi(Q, d, \rho, \mu, D, p) = 0$$

There are six variables and three dimensions, which leads to three Π groups.

Choose Q, d and ρ to involve M, L and T and combine μ, D and p in turn.

Then $$\Pi_1 = \phi_1(Q, d, \rho, \mu) = Q^{a_1}d^{b_1}\rho^{c_1}\mu$$

Equating dimensions,

$$0 = \left[\frac{L^3}{T}\right]^{a_1}[L]^{b_1}\left[\frac{M}{L^3}\right]^{c_1}\left[\frac{M}{LT}\right]$$

Thus, for M: $0 = c_1 + 1$

for L: $0 = 3a_1 + b_1 - 3c_1 - 1$

and for T: $0 = -a_1 - 1$

Hence $a_1 = -1$, $b_1 = 1$ and $c_1 = -1$

so that $$\Pi_1 = \frac{\mu d}{Q\rho}$$

Similarly $$\Pi_2 = \phi_2(Q, d, \rho, D) = Q^{a_2}d^{b_2}\rho^{c_2}D$$

$$\therefore\ 0 = \left[\frac{L^3}{T}\right]^{a_2}[L]^{b_2}\left[\frac{M}{L^3}\right]^{c_2}[L]$$

Thus, for M: $0 = c_2$

for L: $0 = 3a_2 + b_2 - 3c_2 + 1$

and for T: $0 = -a_2$

Hence $a_2 = 0$, $b_2 = -1$ and $c_2 = 0$

so that $$\Pi_2 = \frac{D}{d}$$

$$\Pi_3 = \phi_3(Q, d, \rho, p) = Q^{a_3} d^{b_3} \rho^{c_3} p$$

$$\therefore 0 = \left[\frac{L^3}{T}\right]^{a_3} [L]^{b_3} \left[\frac{M}{L^3}\right]^{c_3} \left[\frac{M}{LT^2}\right]$$

Thus, for M: $0 = c_3 + 1$

for L: $0 = 3a_3 + b_3 - 3c_3 - 1$

and for T: $0 = -a_3 - 2$

Hence $a_3 = -2$, $b_3 = 4$ and $c_3 = -1$

so that $$\Pi_3 = \frac{d^4 p}{Q^2 \rho}$$

Thus $$\phi\left[\frac{\mu d}{Q\rho}; \frac{D}{d}; \frac{d^4 p}{Q^2 \rho}\right] = 0$$

To give the required form, this needs rearrangement (see Art. 8.4). The square root of Π_3 is $\frac{d^2}{Q}\sqrt{\frac{p}{\rho}}$ and dividing this into Π_1 gives $\frac{\mu}{d\sqrt{\rho p}}$ so that

$$\phi\left[\frac{\mu}{d\sqrt{\rho p}}; \frac{D}{d}; \frac{d^2}{Q}\sqrt{\frac{p}{\rho}}\right] = 0$$

or $$\frac{Q}{d^2}\sqrt{\frac{\rho}{p}} = \phi\left[\frac{\mu}{d\sqrt{\rho p}}; \frac{D}{d}\right]$$

from which $$Q = d^2\sqrt{\frac{p}{\rho}}\,\phi\left[\frac{\mu}{d\sqrt{\rho p}}; \frac{D}{d}\right]$$

3. Fluid of density ρ and viscosity μ flows through a circular pipe of diameter d at velocity V. Show that the resistance per unit length R is given by

$$R = \rho V^2 \phi\left[\frac{\rho V d}{\mu}\right]$$

4. Fluid of density ρ and viscosity μ flows at velocity V along a circular pipe of diameter d and length L. Show that the pressure loss due to friction p is given by

$$p = \rho V^2 \phi\left[\frac{\rho V d}{\mu}, \frac{L}{d}\right]$$

5. An orifice of diameter d discharges fluid under a head h in a gravitational field g. The fluid has density ρ and viscosity μ. Show that the quantity flowing Q may be expressed by

$$Q = d^2\sqrt{gh}\,\phi\left[\frac{\mu}{\rho g d^{3/2}}, \frac{h}{d}\right]$$

6. A rectangular weir, breadth B, discharges fluid under a head h in a gravitational field g. The fluid has density ρ, viscosity μ and surface tension T. Show that the quantity flowing per unit width is given by

$$\frac{Q}{B} = kh^{3/2}g^{1/2}\phi\left[\frac{\mu}{h^{3/2}g^{1/2}\rho}, \frac{T}{h^2 g\rho}\right]$$

where k is a constant.

7. Show by dimensional analysis that the flow rate Q over a V-notch may be expressed by

$$Q = g^{1/2}h^{5/2}\phi\left[\frac{g^{1/2}h^{3/2}}{\nu}, \frac{gh^2\rho}{T}, \theta\right]$$

where h is the head over the notch vertex, ρ is the fluid density, ν is the fluid kinematic viscosity, T is the surface tension of the fluid, θ is the angle of the notch and g is the gravitational acceleration.

8. The thrust of a propeller T depends on the propeller diameter D, the speed of rotation N, the speed of advance V, the fluid density ρ and the fluid viscosity μ. Show that the thrust may be expressed by

$$T = \rho D^2 V^2\phi\left[\frac{\mu}{\rho DV}, \frac{ND}{V}\right]$$

9. Assuming that the wavemaking resistance R of a hydrofoil depends on the density of the fluid ρ, the size of the hydrofoil D, the velocity of the hydrofoil V and the gravitational acceleration g, show that

$$R = \rho V^2 D^2\phi\left[\frac{V}{\sqrt{gD}}\right]$$

10. The friction power P dissipated in a water dynamometer is dependent on the brake wheel size D, the brake speed N, the fluid density ρ and the fluid viscosity μ. Show that the power may be expressed by

$$P = D^5 N^3\rho\phi\left[\frac{\mu}{D^2 N\rho}\right]$$

11. Obtain an expression for the resistance R to the motion of a ship, assuming that the important parameters are the size of the ship D, the speed V, the density of the sea ρ, the viscosity of the sea μ and the gravitational acceleration g.

$$\left(\textit{Ans.: } R = \rho L^2 V^2\phi\left[\frac{\rho VD}{\mu}, \frac{V^2}{Dg}\right]\right)$$

9 Dynamic similarity

9.1 Similarity In order to be able to predict the performance of a system from a larger or smaller model, it is necessary to consider the different types of similarity.

(i) *Geometric similarity* requires that the ratio of a length in the system to the corresponding length in the model is the same everywhere.

(ii) *Kinematic similarity* requires that the ratio of velocity and acceleration at any point in the system to the corresponding velocity and acceleration in the model is the same everywhere.

(iii) *Dynamic similarity* requires that the ratio of the force at any point in the system to the corresponding force in the model is the same everywhere.

Geometrical similarity is simple to achieve but it may be shown that in order to obtain kinematic similarity, the conditions of dynamic similarity must also apply. Thus if dynamic similarity is achieved in geometrically similar situations, full similarity results.

9.2 Dynamic similarity If surface tension and other small effects are neglected, the motion of a fluid is governed by a combination of viscous, gravitational and pressure forces. From Newton's Second Law, the vector sum, P, of the external forces acting on a body of mass m causes an acceleration f in the direction of the force, given by

$$P = mf$$

The quantity mf is called the inertia force. Thus for two fluid systems to be dynamically similar, the ratios $\frac{\text{inertia force}}{\text{viscous force}}$, $\frac{\text{inertia force}}{\text{gravitational force}}$ and $\frac{\text{inertia force}}{\text{pressure force}}$ must be equal. Since the inertia force is the sum of the viscous, gravitational and pressure forces, it is only necessary for two of these ratios to be equal to give dynamic similarity.

In fluid mechanics, two problems concerned are submerged bodies (submarines, pipes, pumps, aeroplanes, cars, etc.) where the whole of the surfaces under consideration are in contact with the fluid, and floating bodies, such as ships, which are only partly submerged.

For submerged bodies, gravitational forces are irrevelevant and dynamic similarity is achieved if the ratio $\frac{\text{inertia force}}{\text{viscous force}}$ is equal for each system considered. For floating bodies, however, the ratio of both $\frac{\text{inertia force}}{\text{viscous force}}$ and $\frac{\text{inertia force}}{\text{gravitational force}}$ must be equal for each system.

If V represents velocity, D is a dimension specifying size and t is time, then

$$\text{inertia force} = mf = \rho D^3 \times \frac{V}{t}$$

$$\text{viscous force} = \text{viscous stress} \times \text{area}$$

$$= \tau D^2 = \mu \frac{V}{D} \cdot D^2 = \mu V D$$

and gravitational force $= mg = \rho D^3 g$

The ratio
$$\frac{\text{inertia force}}{\text{viscous force}} = \frac{\rho D^3 \frac{V}{t}}{\mu V D} = \frac{\rho D}{\mu} \cdot \frac{D}{t} = \frac{\rho V D}{\mu}$$

This is the Reynolds number, *Re*.

The ratio
$$\frac{\text{inertia force}}{\text{viscous force}} = \frac{\rho D^3 \frac{V}{t}}{\rho D^3 g} = \frac{V}{tg} = \frac{V}{\frac{D}{V} \cdot g} = \frac{V^2}{Dg}$$

$\frac{V}{\sqrt{Dg}}$ is called the Froude number (Fr), so that $\frac{V^2}{Dg}$ represents $(Fr)^2$. Thus, for geometrically similar submerged bodies, dynamic similarity requires the Reynolds number to be the same for each and for geometrically similar floating bodies, dynamic similarity requires both the Reynolds and Froude numbers to be the same for each.

9.3 Model testing Dimensional analysis provides a method of predicting the performance of systems which are geometrically similar to systems for which the performance is known. This particular application, based on dynamic similarity, has wide uses but is here restricted to problems of resistance to motion of submerged and floating bodies. For this purpose, scale models are built so that any convenient length dimension may be used to specify size.

9.4 Submerged bodies The resistance to motion, R, of a body completely submerged in an incompressible fluid will depend on the size of the body, D, its velocity V and the density ρ and viscosity μ of the liquid. Using the method of dimensional analysis,

$$R = \phi(D, V, \rho, \mu)$$
$$= k(D^a V^b \rho^c \mu^d)$$

Substituting dimensions,

$$\left[\frac{ML}{T^2}\right] = [L]^a \left[\frac{L}{T}\right]^b \left[\frac{M}{L^3}\right]^c \left[\frac{M}{LT}\right]^d$$

Equating powers of M: $1 = c + d$

Equating powers of L: $1 = a + b - 3c - d$

Equating powers of T: $-2 = -b - d$

One solution is $a = 2 - d$, $b = 2 - d$ and $c = 1 - d$

so that
$$R = k(D^{2-d} V^{2-d} \rho^{1-d} \mu^d)$$
$$= k\rho V^2 D^2 \left[\frac{\rho V D}{\mu}\right]^{-d}$$

$\dfrac{\rho VD}{\mu}$ is the Reynolds Number of the flow past the body, Re ,

so that
$$R = \rho V^2 D^2 \phi(Re) \tag{9.1}$$

or
$$\frac{R}{\rho V^2 D^2} = \phi(Re) \tag{9.2}$$

For a series of geometrically similar bodies, dynamic similarity will be achieved if Re is the same for each, when the unknown function ϕ becomes irrelevant. Thus, when Re is the same for each, $R/\rho V^2 D^2$ will be the same for each, from equation (9.2).

If suffix m refers to the model and suffix p refers to the prototype, then for dynamic similarity,

$$(Re)_p \equiv (Re)_m$$

i.e.
$$\left(\frac{\rho VD}{\mu}\right)_p \equiv \left(\frac{\rho VD}{\mu}\right)_m \tag{9.3}$$

It is usual to make model tests in wind tunnels. The tunnel velocity, or *corresponding speed*, is given by

$$V_m = V_p\left(\frac{D_p}{D_m}\right)\left(\frac{\rho_p}{\rho_m}\right)\left(\frac{\mu_m}{\mu_p}\right) \tag{9.4}$$

For these conditions, the resistance of the prototype is given by

$$\left(\frac{R}{\rho V^2 D^2}\right)_p \equiv \left(\frac{R}{\rho V^2 D^2}\right)_m$$

from which
$$R_p = R_m\left(\frac{\rho_p\, V_p^2\, D_p^2}{\rho_m\, V_m^2\, D_m^2}\right) \tag{9.5}$$

Submarines, cars and aeroplanes at low speeds (where compressibility effects may be neglected) are examples of submerged bodies which may be treated by this method.

Pipes may also be considered as submerged bodies. If loss of head is to be predicted,

resistance $R = \rho ghA$ where A is the pipe cross-sectional area

so that the equation $\dfrac{R_p}{R_m} = \dfrac{\rho_p\, V_p^2\, D_p^2}{\rho_m\, V_m^2\, D_m^2}$ becomes

$$\frac{h_p}{h_m} = \frac{V_p^2}{V_m^2} \tag{9.6}$$

This relation is only valid for geometrically similar pipes and if the lengths of pipes considered are not in the correct ratio, it must be modified according to the ratio of the actual pipe length to the geometrically similar length,

i.e.
$$\frac{h_p}{h_m} = \frac{V_p^2}{V_m^2} \times \frac{\text{actual length of pipe}}{\text{geometrically similar length of pipe}}$$

$$= \frac{V_p^2}{V_m^2} \times \frac{L_p}{L_m (D_p/D_m)}$$

$$= \frac{V_p^2 D_m L_p}{V_m^2 D_p L_m} \tag{9.7}$$

If the pipes are of equal length, $L_p = L_m$, so that

$$\frac{h_p}{h_m} = \frac{V_p^2 D_m}{V_m^2 D_p}$$

These results may also be obtained from the Darcy formula,

i.e. $$h_p = \frac{4f_p L_p V_p^2}{2gD_p} \quad \text{and} \quad h_m = \frac{4f_m L_m V_m^2}{2gD_m}$$

At the same Reynolds Number with both pipes of the same material,

$$f_p = f_m$$

so that $$\frac{h_p}{h_m} = \frac{V_p^2 D_m L_p}{V_m^2 D_p L_m}, \quad \text{as before.}$$

9.5 Floating bodies The resistance to motion of the hull of a ship consists of two parts, (*a*) that due to the viscous effects of the water, and (*b*) the wavemaking effects as water is lifted by the hull against gravity. If surface tension effects are ignored, the dimensional analysis of Art. 9.4 may be repeated with the gravitational acceleration g as an additional factor.

$$R = \phi(D, V, \rho, \mu, g) = k(D^a V^b \rho^c \mu^d g^e)$$

Substituting dimensions,

$$\frac{ML}{T^2} = [L]^a \left[\frac{L}{T}\right]^b \left[\frac{M}{L^3}\right]^c \left[\frac{M}{LT}\right]^d \left[\frac{L}{T^2}\right]^e$$

Equating powers of M: $1 = c + d$

Equating powers of L: $1 = a + b - 3c - d + e$

Equating powers of T: $-2 = -b - d - 2e$

One solution is $a = 2 - d + e$, $b = 2 - d - 2e$ and $c = 1 - d$.

so that $$R = k(D^{2-d+e} V^{2-d-2e} \rho^{1-d} \mu^d g^e)$$

$$= k\rho V^2 D^2 \left[\frac{\rho V D}{\mu}\right]^{-d} \left[\frac{V^2}{Dg}\right]^{-e}$$

$$= k\rho V^2 D^2 \left[\frac{\rho V D}{\mu}\right]^{-d} \left[\frac{V}{\sqrt{Dg}}\right]^{-2e}$$

$\dfrac{V}{\sqrt{Dg}}$ is the Froude Number of the flow, Fr, so that

$$R = \rho V^2 D^2 \phi(Re, Fr) \tag{9.8}$$

or

$$\frac{R}{\rho V^2 D^2} = \phi(Re, Fr)$$

For dynamic similarity, Re must be the same for the model and prototype and Fr must be the same for the model and prototype. Model tests are conducted in water tanks so that the fluid properties are substantially the same for the model and prototype.

Then, for equality of Reynolds Numbers,

$$V_m = V_p \left(\frac{D_p}{D_m}\right)$$

and for equality of Froude Numbers,

$$V_m = V_p \sqrt{\frac{D_m}{D_p}}$$

These two conditions are incompatible and so dynamic similarity cannot be achieved.

If R_v is the viscous resistance and R_w is the wavemaking resistance,

$$R = R_v + R_w$$

The viscous resistance is approximately proportional to the wetted area and the square of the velocity, so that

$$R_v = \tfrac{1}{2} \rho C_D A V^2 \tag{9.9}$$

where C_D is a dimensionless drag coefficient.

The value of C_D is determined by experiment on a flat plate of the same material and surface finish, so that for the model,

$$R_{v_m} = \tfrac{1}{2} \rho C_D A_m V_m^2$$

and for the prototype,

$$R_{v_p} = \tfrac{1}{2} \rho C_D A_p V_p^2$$

The wavemaking resistance is given by equation (9.8) when μ is omitted from the analysis,

i.e.

$$R_w = \rho V^2 D^2 \phi(Fr)$$

Tests are then conducted at the corresponding speed given by the Froude equality,

i.e.

$$V_m = V_p \sqrt{\frac{D_m}{D_p}} \tag{9.10}$$

so that $$\left(\frac{R_w}{\rho V^2 D^2}\right)_m = \left(\frac{R_w}{\rho V^2 D^2}\right)_p$$

or $$\frac{R_{w_p}}{R_{w_m}} = \frac{\rho_p V_p^2 D_p^2}{\rho_m V_m^2 D_m^2} = \frac{\rho_p D_p^3}{\rho_m D_m^3} \qquad (9.11)$$

The total resistance, R_p, is then obtained from

$$\frac{R_p - R_{v_p}}{R_m - R_{v_m}} = \frac{\rho_p D_p^3}{\rho_m D_m^3} \qquad (9.12)$$

where R_m is the total model resistance measured in the test at the speed given by equation (9.10).

In the case of a hydrofoil, which is considered to have no viscous resistance, the resistance is given directly by equation (9.11).

9.6 Drag and lift coefficients Submerged bodies are so common that it is convenient to express the resistance parallel to the fluid flow as a *drag force* and to use a drag coefficient C_D, as in Art. 9.5, to determine this force, given by

$$R_D = \tfrac{1}{2}\rho C_D A V^2 \qquad (9.13)$$

But from equation (9.1),

$$R = \rho V^2 D^2 \phi(Re)$$

so that C_D is a function of Re. Thus, if C_D is found for a model, it will apply also to a geometrically similar system under dynamically similar conditions.

In a similar way, the lift force, R_L, normal to the direction of motion may be expressed in terms of a lift coefficient C_L, given by

$$R_L = \tfrac{1}{2}\rho C_L A V^2 \qquad (9.14)$$

C_L is also a function of Re and so has the same value for geometrically similar systems under dynamically similar conditions.

The value of A in equations (9.13) and (9.14), equivalent to the D^2 term in equation (9.1), is chosen by convention for common applications, i.e. wing plan area for aircraft and frontal area for cars.

Equations (9.13) and (9.14) may be rearranged to give

$$C_D = \frac{R_D/A}{\tfrac{1}{2}\rho V^2} \qquad \text{and} \qquad C_L = \frac{R_L/A}{\tfrac{1}{2}\rho V^2}$$

The term $\tfrac{1}{2}\rho V^2$ is called the dynamic pressure of the fluid at velocity V and represents, from Art. 4.7, the increase in pressure when the fluid is brought to rest. Thus, at the *stagnation point* S on the submerged body shown in Fig. 9.1, the free stream pressure is increased by $\tfrac{1}{2}\rho V^2$. Elsewhere on the surface, the free stream pressure will be reduced or increased, depending on whether the local velocity is greater than, or less than, the free stream velocity.

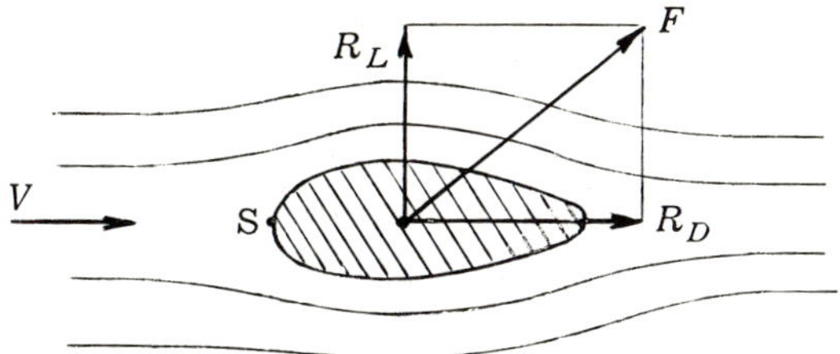

Fig. 9.1

The resultant force, F, on the body, which might be a turbine blade, aerofoil or propeller, may be resolved into the drag force R_D and lift force R_L, respectively parallel and normal to the direction of flow, to enable the performance of the body to be conveniently analysed.

9.7 Model testing problems It is sometimes found that the achievement of dynamic similarity is only possible if the fluid or its density is changed; this is to avoid excessive corresponding speed requirements when testing small models.

If the performance of an aircraft, to fly at 250 m/s, is examined by means of a ⅕ scale model, tested in air at the same pressure and temperature as the real aircraft operating conditions, then ρ and μ are the same for the model and prototype.

For dynamic similarity,

$$\left(\frac{\rho V D}{\mu}\right)_m = \left(\frac{\rho V D}{\mu}\right)_p$$

But $\rho_m = \rho_p$ and $\mu_m = \mu_p$, so that $V_m D_m = V_p D_p$. Since $D_p = 5D_m$, then $V_m = 5V_p$ and so the model test needs to be conducted at 1 250 m/s. At such high speeds, other effects occur which would render the experimental results useless and so, to avoid this problem, the ambient pressure in model tests may be increased to increase ρ_m and give a smaller value for V_m. The tests may also be conducted in an entirely different fluid of greater density.

9.8 Experimental results Dimensional analysis will not produce a complete answer to a problem and experiments are necessary to supply further data. In order to use such experimental results for all geometrically similar systems, they must be presented graphically in non-dimensional form.

Suppose that for a particular problem, dimensional analysis shows that

$$\Pi_1 = \phi(\Pi_2, \Pi_3, \Pi_4)$$

where Π_1, Π_2, Π_3 and Π_4 are non-dimensional groups.

Experiments are performed so that the values of these groups can be determined. The results are then displayed as graphs of Π_1 against Π_2 with Π_4 constant and Π_3 as contour, as shown in Fig. 9.2. To produce such a set of results would need a considerable number of tests but the graphs are then available to predict the performance of any geometrically similar system. Such graphs would be used for a range of pumps or propellers.

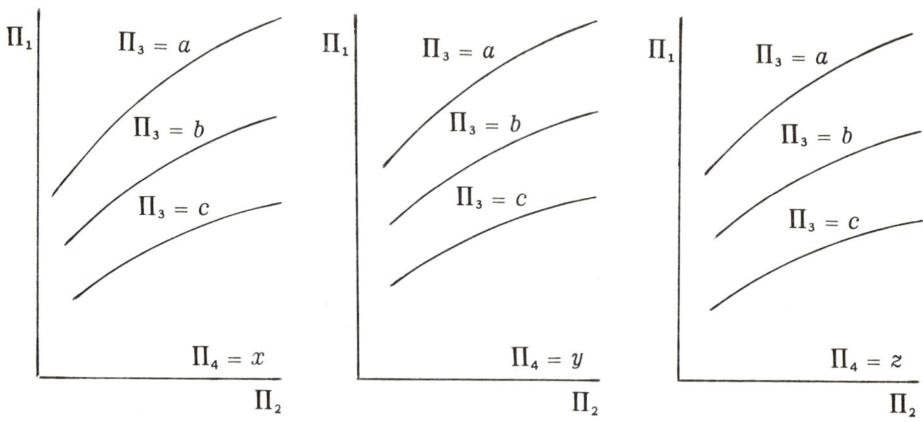

Fig. 9.2

1. *The power requirements of a small submarine cruising submerged in sea water at a speed of 15 knots are to be determined using a 1/10 th scale model in a high density wind tunnel. The model drag is measured as 300 N when the tunnel pressure is 6 bar and the tunnel temperature is 27°C.*

Determine the tunnel velocity and the power required for the submarine.

For sea water, $\rho = 1\,030\ kg/m^3$ *and* $\mu = 1{\cdot}12 \times 10^{-3}\ N\,s/m^2$.
For air at 27°C, 6 bar, $\rho = 7{\cdot}06\ kg/m^3$ *and* $\mu = 18{\cdot}46 \times 10^{-6}\ N\,s/m^2$.
1 knot = 6 080 ft/h = 0·514 m/s.

From equation (9.4), the corresponding speed of the model,

$$V_m = V_p\left(\frac{D_p\,\rho_p\,\mu_m}{D_m\,\rho_m\,\mu_p}\right)$$

$$= 15 \times 0{\cdot}514\left(\frac{10 \times 1\,030 \times 18{\cdot}46 \times 10^{-6}}{1 \times 7{\cdot}06 \times 1{\cdot}12 \times 10^{-3}}\right)$$

$$= \underline{186\ \text{m/s}}$$

At the corresponding speed, $$R_p = R_m\left(\frac{\rho_p\,V_p^{\,2}\,D_p^{\,2}}{\rho_m\,V_m^{\,2}\,D_m^{\,2}}\right)$$

$$= 300\left(\frac{1\,030 \times [15 \times 0{\cdot}514]^2 \times 10^2}{7{\cdot}06 \times 186^2 \times 1^2}\right)$$

$$= 7\,600\ \text{N}$$

Power required to overcome resistance

$$= 7\,600 \times (15 \times 0{\cdot}514)$$

$$= 58\,700\ \text{W} \quad \text{or} \quad \underline{58{\cdot}7\ \text{kW}}$$

2. *A scale model of a ship is 4 m long and is towed through a water tank at 2·3 m/s. The wetted area of the ship is 500 m² and its length is 50 m. What is the power requirement of the ship, if the model resistance at the corresponding speed is 56 N?*

For both model and ship, the drag friction coefficient C_D is 0·002 1. It may be assumed that the water in the tank has the same properties as the water in which the ship operates, which has a density of 1 026 kg/m³.

For a floating body, corresponding speeds are obtained from equality of Froude Numbers,

$$V_p = V_m \sqrt{\frac{D_p}{D_m}} \quad . \quad . \quad . \quad . \quad \text{from equation (9.10)}$$

$$= 2{\cdot}3 \sqrt{\frac{50}{4}} = 8{\cdot}14 \text{ m/s}$$

$$\text{Surface area of model} = 500 \times \left(\frac{4}{50}\right)^2 = 3{\cdot}2 \text{ m}^2$$

$$\text{Viscous drag, } R_v = \tfrac{1}{2} \rho C_D A V^2 \quad . \quad . \quad . \quad \text{from equation (9.9)}$$

$$\text{At } 8{\cdot}14 \text{ m/s,} \quad R_{v_p} = \tfrac{1}{2} \times 1\,026 \times 0{\cdot}002\,1 \times 500 \times 8{\cdot}14^2 = 35\,600 \text{ N}$$

$$\text{At } 2{\cdot}5 \text{ m/s,} \quad R_{v_m} = \tfrac{1}{2} \times 1\,026 \times 0{\cdot}002\,1 \times 3{\cdot}2 \times 2{\cdot}3^2 = 18{\cdot}2 \text{ N}$$

Thus, from equation (9.12)

$$\frac{R_p - R_{v_p}}{R_m - R_{v_m}} = \frac{\rho_p D_p^3}{\rho_m D_m^3}$$

i.e.

$$\frac{R_p - 35\,600}{56 - 18{\cdot}2} = \frac{1\,026 \times 50^3}{1\,026 \times 4^3}$$

from which $R_p = 109\,500$ N

$$\text{Therefore power required} = 109\,500 \times 8{\cdot}14$$

$$= 893\,000 \text{ W} \quad \text{or} \quad \underline{893 \text{ kW}}$$

3. *A pipe 50 m long and 20 mm diameter carries water at 10°C. The water flows with a velocity of 5 m/s and the loss of head in the pipe is 60 m. A pipe of similar material is to carry oil at the corresponding speed, the oil having a density of 850 kg/m³ and a viscosity of 0·032 kg/m s. Calculate the pressure difference between the ends of the oil pipe, which is 75 m long and 1·5 m diameter.*

For water at 10°C, $\rho = 10^3$ kg/m³ and $\mu = 1{\cdot}3 \times 10^{-3}$ kg/m s.

At the corresponding speeds, the Reynolds numbers must be equal,

i.e.

$$\left(\frac{\rho V D}{\mu}\right)_{\text{water}} = \left(\frac{\rho V D}{\mu}\right)_{\text{oil}}$$

i.e.
$$\frac{1\,000 \times 5 \times 0{\cdot}02}{1{\cdot}3 \times 10^{-3}} = \frac{850 \times V_{oil} \times 1{\cdot}5}{0{\cdot}032}$$

$$\therefore \quad V_{oil} = 1{\cdot}93 \text{ m/s}$$

From equation (9.7),
$$\frac{h_{oil}}{h_{water}} = \frac{V_{oil}^2 \, D_{water} \, L_{oil}}{V_{water}^2 \, D_{oil} \, L_{water}}$$

$$\therefore \quad h_{oil} = 60 \times \frac{1{\cdot}93^2 \times 0{\cdot}02 \times 75}{5^2 \times 1{\cdot}5 \times 50}$$

$$= 0{\cdot}178 \text{ m}$$

$$\therefore \quad p = \rho g h$$

$$= 850 \times 9{\cdot}81 \times 0{\cdot}178 = \underline{1480 \text{ N/m}^2}$$

4. In order to determine the drag of an aircraft designed to fly at 150 m/s, a $\frac{1}{25}$ scale model is tested in a wind tunnel at 29 atmospheres. The drag of the model at the corresponding speed is 180 N. Assuming that the viscosity of air is independent of pressure and that the tunnel temperature is equal to the air temperature at which the aircraft will operate, determine the power required by the aircraft.

(*Ans.*: 783 kW)

5. A $\frac{1}{5}$ scale model of an automobile component with frontal area 100 cm² is tested in water to determine its drag. The automobile moves at 18 m/s and at the corresponding speed in the water test, the model drag was 7 N. Calculate the power required to overcome the component drag and determine the component drag coefficient.
For air, $\rho = 1{\cdot}412$ kg/m³ and $\mu = 1{\cdot}599 \times 10^{-5}$ N s/m².
For water, $\rho = 1\,000$ kg/m³ and $\mu = 1\,002 \times 10^{-6}$ N s/m². (*Ans.*: 22·7 W, 0·55)

6. A hydrofoil craft is to travel at 30 m/s in sea water. The foils are 10 m in length and in tests conducted at the corresponding speed in a fresh water tank, the scale model 0·5 m long had a resistance of 5 N. Determine the power required for the hydrofoil craft.
For sea water, $\rho = 1\,030$ kg/m³ and for fresh water, $\rho = 1\,000$ kg/m³.

(*Ans.*: 1 236 kW)

7. A ship is to be built 150 m in length with a wetted surface area of 3 000 m². The ship is to operate at 20 knots. A $\frac{1}{40}$ scale model is towed at the corresponding speed in a tank of fresh water and the model resistance is 19 N. For both model and ship, the flat plate drag coefficient may be assumed to be 0·002 5 (based on S.I. units). Determine the power requirement for the ship.
For sea water, $\rho = 1\,026$ kg/m³ and $\mu = 11{\cdot}2 \times 10^{-4}$ N s/m².
For fresh water, $\rho = 1\,000$ kg/m³ and $\mu = 11{\cdot}5 \times 10^{-4}$ N s/m².
1 knot = 0·514 m/s. (*Ans.*: 12 750 kW)

8. Determine the power required to drive a ship 100 m in length with a wetted area of 1 300 m² at 11 m/s in fresh water. The resistance in fresh water at the same temperature of a $\frac{1}{10}$ scale model run at corresponding speed is 500 N. The drag coefficient for flat plates may be obtained from the relation $C_D = 0{\cdot}074/Re^{0{\cdot}2}$.
For fresh water, $\rho = 1\,000$ kg/m³ and $\mu = 1\,136 \times 10^{-6}$ N s/m².

(*Ans.*: 4 560 kW)

9. The velocity of flow of water through an 80 mm diameter pipe is 5 m/s and the loss of head at this speed is 48 m of water per 100 m of pipe. A pipe of similar material, 400 mm diameter, carries air at 300 K at the corresponding speed. What will be the loss of head, in metres of water, per 100 m of pipe?
For air at 300 K, $\mu = 18{\cdot}46 \times 10^{-6}$ N s/m², $\rho = 1{\cdot}177$ kg/m³.
For water, $\mu = 1{\cdot}136 \times 10^{-6}$ N s/m², $\rho = 1\,000$ kg/m³. (*Ans.*: 0·088 m)

10 Turbomachinery

10.1 Introduction In a hydraulic turbomachine, liquid flows over curved blades, or through passages between curved blades, on a rotor. The force exerted on the blades derives from the rate of change of momentum of the liquid and this in turn produces a torque on the rotor shaft. The power developed is then the product of torque and angular speed.

When power is produced by a turbomachine, it is called a turbine and when power is absorbed to raise pressure, it is called a pump. Similar machines exist for compressible fluids but these are beyond the scope of this book.

10.2 Dimensional analysis applied to turbomachinery In turbomachines, the flow is extremely turbulent and viscous effects are small. Thus viscosity and Reynolds number are absent in dimensional analysis applied to turbomachines and the relevant variables are pressure (p), shaft power (P), efficiency (η), volume flow rate (Q), fluid density (ρ) and runner diameter (D). Pressure may be expressed in terms of the head ($p = \rho g H$) and gH is preferred as a variable.

These variables are not all independent and it is usually considered that gH, P and η are dependent on the remainder.

$$gH = f_1(Q, N, \rho, D) \qquad (10.1)$$

$$P = f_2(Q, N, \rho, D) \qquad (10.2)$$

and

$$\eta = f_3(Q, N, \rho, D) \qquad (10.3)$$

From equation (10.1),

$$gH = k(Q^a N^b \rho^c D^d)$$

Substituting dimensions,

$$\left[\frac{L^2}{T^2}\right] = \left[\frac{L^3}{T}\right]^a \left[\frac{1}{T}\right]^b \left[\frac{M}{L^3}\right]^c [L]^d$$

Equating powers of M: $0 = c$

Equating powers of L: $2 = 3a - 3c + d$

Equating powers of T: $-2 = -a - b$

Three indices may be eliminated from these equations and one solution is $b = 2 - a$, $c = 0$ and $d = 2 - 3a$,

so that

$$gH = k(Q^a N^{2-a} D^{2-3a}) = kN^2D^2\left[\frac{Q}{ND^3}\right]^a$$

or
$$\frac{gH}{N^2D^2} = \phi\left[\frac{Q}{ND^3}\right] \tag{10.4}$$

Similarly, from equations (10.2) and (10.3), it is found that

$$\frac{P}{\rho N^3D^5} = \phi\left[\frac{Q}{ND^3}\right] \tag{10.5}$$

and
$$\eta = \phi\left[\frac{Q}{ND^3}\right] \tag{10.6}$$

For a range of geometrically similar turbines or pumps, dynamic similarity is achieved if $\frac{Q}{ND^3}$ is the same for each. At this condition, the values of $\frac{gH}{N^2D^2}$, $\frac{P}{\rho N^3D^5}$ and η will also be the same for each machine in the range. These parameters are found by experiment for a particular design of machine *at the point of maximum efficiency*; these will be the same for all similar machines and are given the following names:

Flow coefficient, or discharge number, $C_Q = \frac{Q}{ND^3}$ (10.7)

Head coefficient, $C_H = \frac{gH}{N^2D^2}$ (10.8)

Power coefficient, $C_P = \frac{P}{\rho N^3D^5}$ (10.9)

These non-dimensional parameters may be used to plot the performance characteristics of a turbomachine, which will apply to all geometrically similar machines. The relation between these parameters is shown in Fig. 10.1.

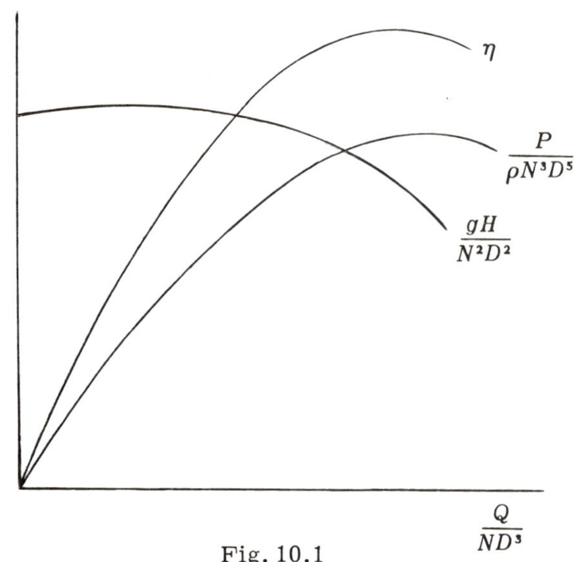

Fig. 10.1

Equations (10.4), (10.5) and (10.6) show that in any range of machines operating at corresponding points, doubling the speed will double the discharge, quadruple the head and multiply the power eight-fold.

10.3 Specific speed The specific speed of a turbomachine is the speed at which a geometrically similar machine would need to run to produce unit output from unit input at maximum efficiency. This speed has a typical value for each different design of machine and is a useful parameter for selecting the optimum type for given performance requirements.

For dynamic similarity, the dimensionless groups $\frac{gH}{N^2D^2}$, $\frac{Q}{ND^3}$ and $\frac{P}{\rho N^3D^5}$ must be the same for each machine and, as stated in Art. 8.4, these groups may be multiplied or divided amongst themselves to produce other dimensionless groups, as required. These new groups must also be the same for dynamically similar machines.

For a *turbine*, the specific speed is defined as the speed at which a similar turbine would generate an output of 1 kW under a head of 1 m. This is determined by eliminating the variable D from the groups $\frac{gH}{N^2D^2}$ and $\frac{P}{\rho N^3D^5}$, i.e. between C_H and C_P.

Thus
$$\frac{\left(\frac{P}{\rho N^3D^5}\right)^{1/2}}{\left(\frac{gH}{N^2D^2}\right)^{5/4}} = \frac{NP^{1/2}}{\rho^{1/2}g^{5/4}H^{5/4}}$$

This ratio must be the same for the actual turbine and the specific turbine. Denoting the variables for the specific turbine by suffix s,

$$\frac{N_s P_s^{1/2}}{\rho^{1/2}g^{5/4}H_s^{5/4}} = \frac{NP^{1/2}}{\rho^{1/2}g^{5/4}H^{5/4}}$$

But $P_s = 1$ kW and $H_s = 1$ m and so, assuming ρ to be the same for each,

$$N_s = \frac{NP^{1/2}}{H^{5/4}} \qquad (10.10)$$

For a *pump*, the specific speed is defined as the speed at which a similar pump would deliver an output of 1 m³/s against a head of 1 m. This is determined by eliminating the variable D from the groups $\frac{gH}{N^2D^2}$ and $\frac{Q}{ND^3}$, i.e. between C_H and C_Q.

Thus
$$\frac{\left(\frac{Q}{ND^3}\right)^{1/2}}{\left(\frac{gH}{N^2D^2}\right)^{3/4}} = \frac{NQ^{1/2}}{g^{3/4}H^{3/4}}$$

This ratio must be the same for the actual pump and the specific pump.

Denoting the variables for the specific pump by suffix s,

$$\frac{N_s Q_s^{1/2}}{g^{3/4} H_s^{3/4}} = \frac{N Q^{1/2}}{g^{3/4} H^{3/4}}$$

But $Q_s = 1\ \text{m}^3/\text{s}$ and $H_s = 1$ m, so that

$$N_s = \frac{N Q^{1/2}}{H^{3/4}} \tag{10.11}$$

The specific speeds for turbines and pumps depend on the systems of units employed and would have different values if power, discharge and head were expressed in horsepower and feet units rather than S.I. Units. *

10.4 Choice of machine The specific speed of a turbomachine depends only on the shape and might be termed a 'shape factor'. The value is determined from coefficients defined at the point of maximum efficiency and thus the *optimum* choice of a machine for particular performance requirements is restricted. The specific speed for these requirements is obtained from equations (10.10) or (10.11) and this then determines the most suitable type of machine, the only other variable being the size necessary to provide the required output.

The 'shape' of a turbine or pump is determined by two factors: the degree of reaction and the flow direction.

10.5 Degree of reaction When a liquid passes through a machine, both the kinetic and pressure energy of the liquid may change; the distribution of the the change between kinetic and pressure energy defines the *degree of reaction*. When there is no change in static pressure across the runner, the degree of reaction is zero and the machine is an impulse design. There are no impulse pumps but impulse turbines, known as Pelton wheels, are used (see Art. 5.4).

In an impulse turbine, the available head is converted to kinetic energy through fixed nozzles and the high velocity stream of water passes over blades, or buckets, where the change of momentum produces the power. The runner is not full of water; when the water falls from a bucket, it remains empty until it meets the next jet and the pressure is atmospheric throughout. The specific speed of a Pelton wheel is low ($N_s < 40$) and this design is used when a large head of water is available.

Turbines and pumps in which there is a change of kinetic and pressure energy in the runner have a degree of reaction and are called reaction machines. The degree of reaction is determined by the shape of the runner and the blade angles. All such machines *must run full of water* in order to achieve the static pressure change.

Reaction turbines include Francis and Kaplan (or propeller) types which offer a range of specific speeds. Pumps include centrifugal, axial and mixed flow designs which again offer a range of specific speeds. These various types are described briefly in the following articles, while centrifugal pumps are considered in more detail in Chapter 11.

* Although specific speeds are normally quoted in rev/min, the units appropriate to S.I. definitions are actually $\text{rev min}^{-1}\text{kW}^{1/2}\text{m}^{-5/4}$ for turbines and $\text{rev min}^{-1}\text{s}^{-1/2}\text{m}^{3/4}$ for pumps.

10.6 Reaction turbines The arrangement of a turbine plant is shown in Fig. 10.2. Water is supplied to the turbine from a reservoir and thence, via a draft tube, to a tail race. The object of the draft tube is to keep the turbine full of water and it is divergent to reduce the final velocity of the water, thus keeping the loss of kinetic energy at exit, $\frac{V_{exit}^2}{2g}$ to a minimum. The height of the draft tube, Z, is limited by the need to keep the outlet pressure, p_2, at the runner above about 2·5 m of water absolute, otherwise bubbles of vapour will be released which will damage the turbine. This phenomenon is known as cavitation.

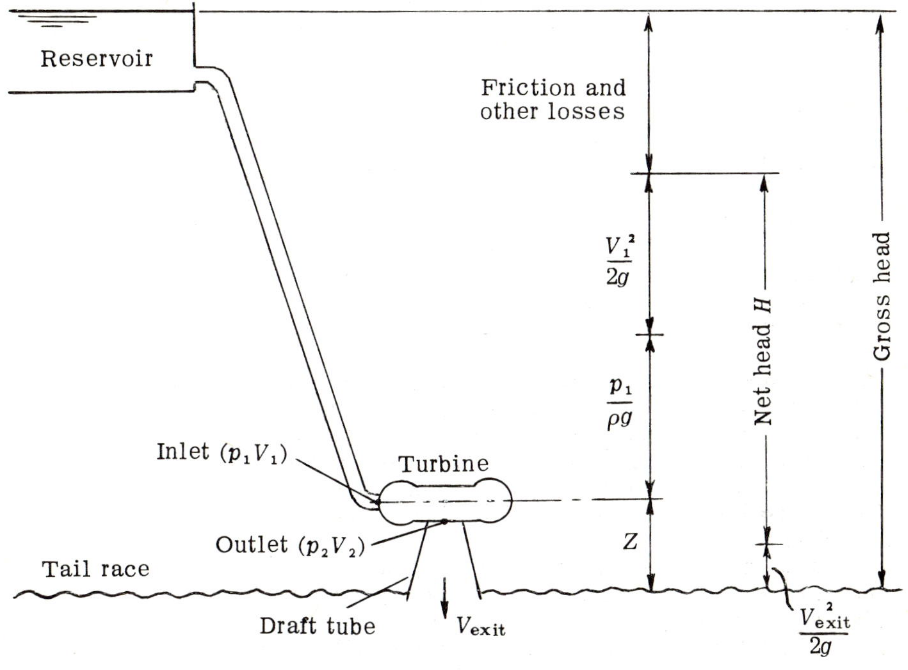

Fig. 10.2

Some of the gross head is lost due to pipe friction, entry loss, bends, etc., and some is loss in kinetic energy at exit; the remaining head is available to produce power.

Thus net head, H = gross head − losses − exit K.E.

10.7 Types of reaction turbine

(*a*) *Radial flow turbine* The principle of the Francis radial flow turbine is shown in Fig. 10.3. Water enters a spiral volute chamber and then flows radially through stationary pivoted guide vanes; these direct the water, ideally without shock, on to the moving vanes attached to the runner. The water flows radially inwards through the vanes but leaves the runner axially to enter the draft tube. Francis turbines have medium specific speeds (N_s = 40 − 350) and are suited to heads between 20 and 200 m.

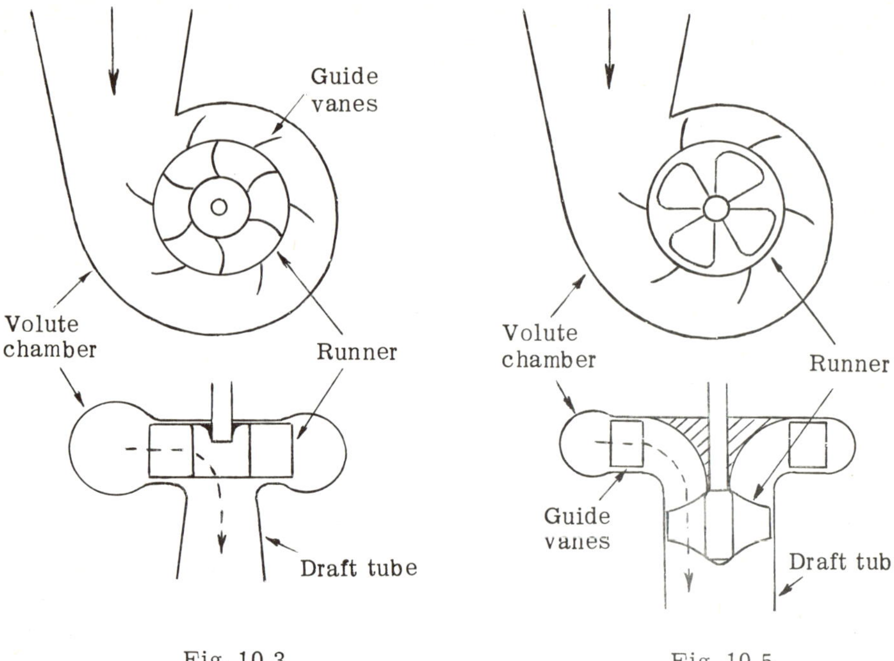

Fig. 10.3 Fig. 10.5

(*b*) *Mixed flow turbines* The runner is generally similar to that of the radial flow turbine but is designed to give partly radial and partly axial flow. Fig. 10.4 shows variations in the flow direction for different runners, together with typical specific speeds.

It will be seen that the specific speed increases as the flow becomes more axial and less radial.

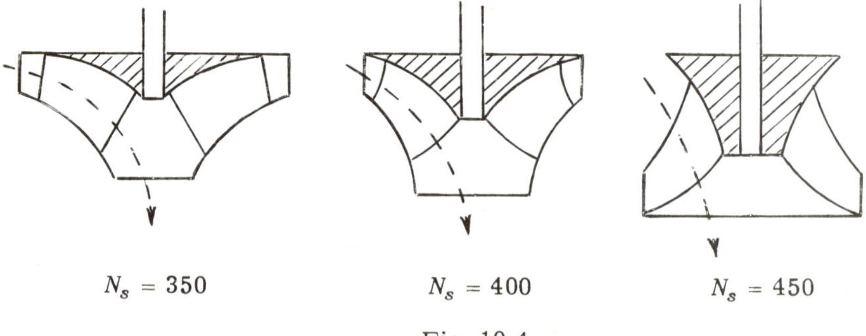

Fig. 10.4

(*c*) *Axial flow turbine* In a Kaplan axial flow or propeller turbine, Fig. 10.5, the water enters a spiral volute chamber and then flows radially through stationary pivoted guide vanes. It is then turned into the axial direction *before* passing through the runner, which is similar to a propeller.

Kaplan turbines have a high specific speed (N_s = 430 - 750) and are suited to low net heads, i.e. between 3 and 25 m.

10.8 Types of pump

(*a*) *Radial flow pump* The principle of a radial flow or centrifugal pump is shown in Fig. 10.6. Water enters the impeller (or rotor) eye axially and then flows radially outwards through the blade passages. The specific speed of a centrifugal pump may be up to 100 but when $N_s < 30$, a set of diffuser blades is fixed round the rotor to improve efficiency.

Centrifugal pumps give relatively high heads with a low flow rate.

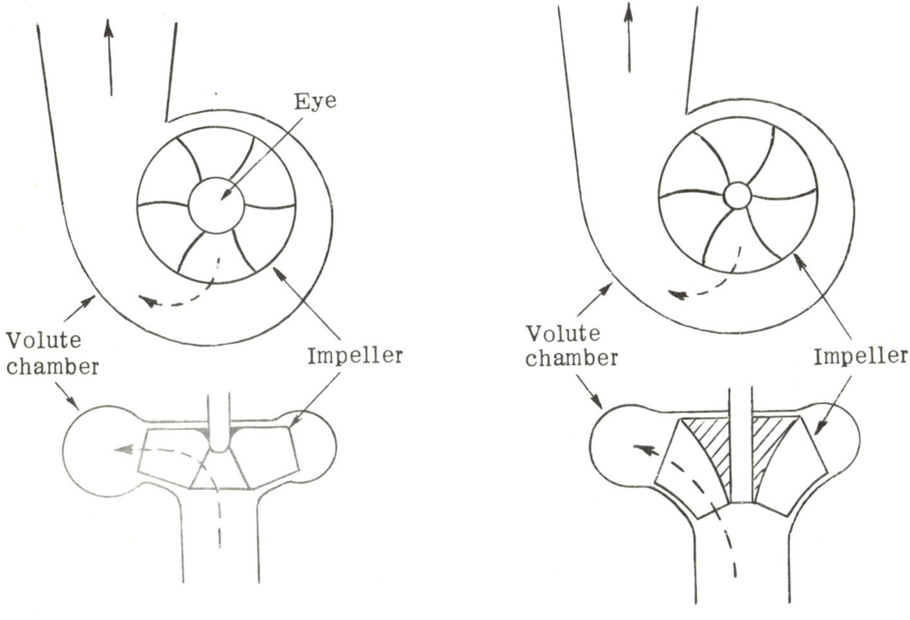

Fig. 10.6

Fig. 10.7

(*b*) *Mixed flow pumps* In a mixed flow pump, Fig. 10.7, the water enters axially and then passes through an impeller which gives it a partly radial flow at exit. As with turbines, the transition from axial to radial flow is gradual and the specific speed and general performance lie between those for centrifugal and axial flow pumps. Specific speeds range from 100 to 200.

(*c*) *Axial flow, or propeller, pump* The principle of action is shown in Fig. 10.8. Water flows axially through a propeller-type runner and then through stationary guide vanes.

The specific speed of a propeller pump is high ($N_s > 200$) and this design is suitable for cases where a high flow rate with a low head is required.

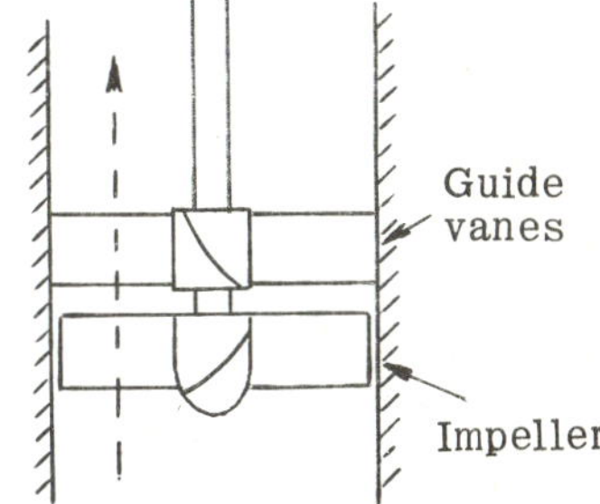

Fig. 10.8

10.9 Machine choice based on specific speed It has been seen that the optimum choice of machine type for a given performance is dictated by its specific speed and this information is summarized in Figs. 10.9 and 10.10, which show the range of specific speeds for different types of turbine and pump respectively.

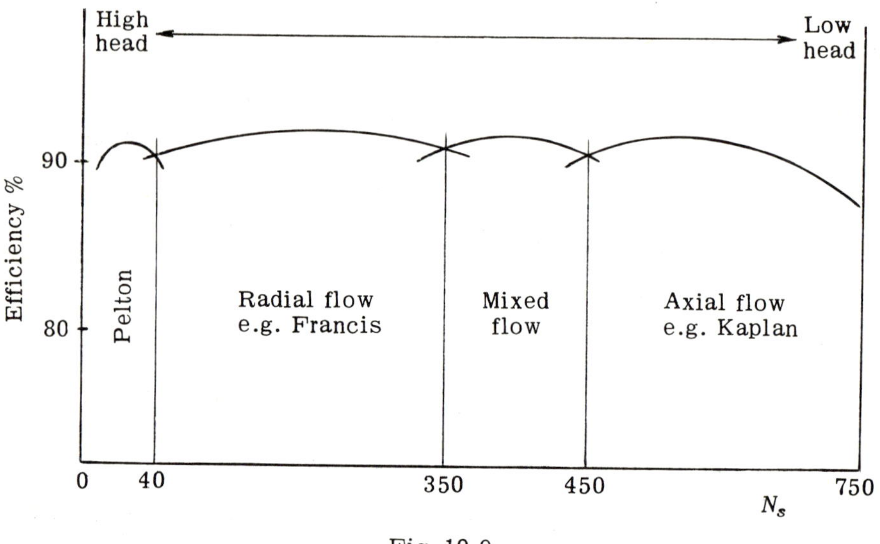

Fig. 10.9

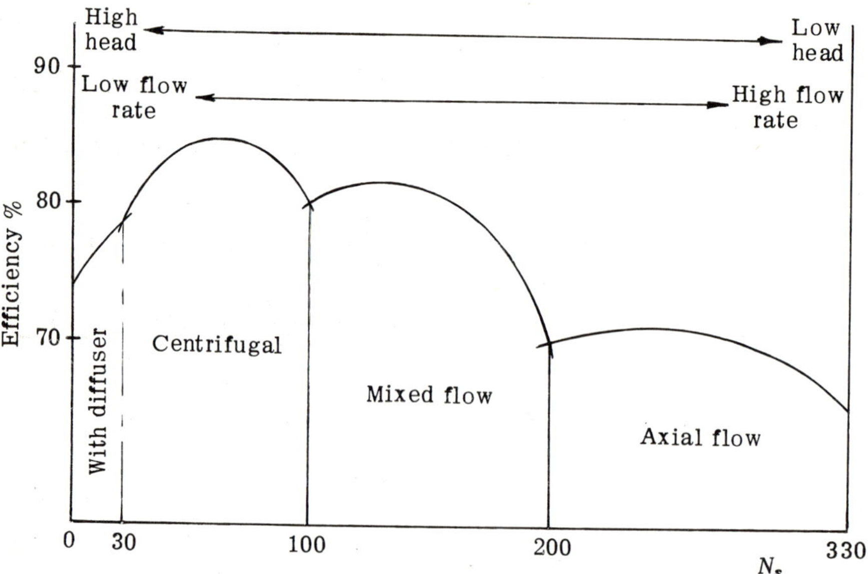

Fig. 10.10

1. *A Francis turbine is required to develop about 4 500 kW at maximum efficiency when running at 110 rev/min with a head of 14 m. A scale model is tested with a head of 3·7 m and at maximum efficiency, the speed was 210 rev/min, the flow rate was 1·45 m³/s and the power output was 44·25 kW.*

Determine the ratio of sizes of turbine and model, the specific speed, the flow rate required through the turbine and its power output and efficiency.

From Art. 10.3, the head coefficient C_H is the same for the turbine and model at the point of maximum efficiency. Denoting model by suffix m and turbine by suffix t,

$$C_H = \left(\frac{gH}{N^2D^2}\right)_m = \left(\frac{gH}{N^2D^2}\right)_t \quad \text{. . from equation (10.8)}$$

so that $$\frac{D_t}{D_m} = \sqrt{\frac{H_t \times N_t^2}{H_m \times N_m^2}} = \sqrt{\frac{14 \times 210^2}{3{\cdot}7 \times 110^2}} = \underline{3{\cdot}714}$$

From the results of the model test,

$$N_s = \frac{NP^{1/2}}{H^{5/4}} \quad \text{. . . . . . . from equation (10.10)}$$

$$= \frac{210 \times 44{\cdot}25^{1/2}}{3{\cdot}7^{5/4}} = \underline{272{\cdot}2 \text{ rev/min}}$$

The flow coefficient C_Q is also the same for model and turbine at the point of maximum efficiency,

i.e. $$C_Q = \left(\frac{Q}{ND^3}\right)_m = \left(\frac{Q}{ND^3}\right)_t \quad \text{. . from equation (10.7)}$$

so that $$Q_t = Q_m \times \frac{N_t D_t^3}{N_m D_m^3} = 1{\cdot}45 \times \frac{110}{210} \times 3{\cdot}714^3 = \underline{38{\cdot}94 \text{ m}^3/\text{s}}$$

Similarly, the power coefficient C_P is the same for model and turbine at the point of maximum efficiency,

i.e. $$C_P = \left(\frac{P}{\rho N^3D^5}\right)_m = \left(\frac{P}{\rho N^3D^5}\right)_t \quad \text{. from equation (10.9)}$$

so that $$P_t = P_m \times \frac{N_t^3 D_t^5}{N_m^3 D_m^5} = 44{\cdot}25 \times \left(\frac{110}{210}\right)^3 \times 3{\cdot}714^5 = \underline{4\,490 \text{ kW}}$$

Alternatively, $$N_s = \frac{N_t P_t^{1/2}}{H_t^{5/4}}$$

i.e. $$272{\cdot}2 = \frac{110 P^{1/2}}{14^{5/4}}$$

from which $$P = \underline{4\,490 \text{ kW}}$$

The efficiency is given by

$$\eta = \frac{\text{power output}}{\text{power input}} = \frac{P}{\rho g QH}$$

$$= \frac{4\,490 \times 10^3}{10^3 \times 9{\cdot}81 \times 38{\cdot}94 \times 14} = 0{\cdot}839\,6 \quad \text{or} \quad \underline{83{\cdot}96\%}$$

2. *A small rotary pump with an impeller 0·2 m diameter runs at maximum efficiency and discharges 0·065 m³/s of water against a head of 4·27 m when running at 1 800 rev/min. What is the specific speed?*

A similar pump is to deliver 0·92 m³/s of water to a condenser. If the impeller diameter is 0·4 m, at what speed should this pump run and what will be the head produced?

From equation (10.11), $N_s = \dfrac{NQ^{1/2}}{H^{3/4}}$

$$= \frac{1\,800 \times 0{\cdot}065^{1/2}}{4{\cdot}27^{3/4}} = \underline{154{\cdot}5 \text{ rev/min}}$$

If both pumps are to operate at maximum efficiency, the flow coefficient C_Q will be the same for each.

Denoting the small pump by suffix 1 and the large pump by suffix 2,

$$\frac{Q_1}{N_1 D_1^3} = \frac{Q_2}{N_2 D_2^3} \quad . \quad . \quad . \quad \text{from equation (10.7)}$$

$$\therefore \; N_2 = N_1 \frac{Q_2}{Q_1}\left(\frac{D_1}{D_2}\right)^3$$

$$= 1\,800 \times \frac{0{\cdot}92}{0{\cdot}065} \times \left(\frac{0{\cdot}2}{0{\cdot}4}\right)^3 = 3\,185 \text{ rev/min}$$

Similarly, the head coefficient C_H will be the same for each,

i.e.
$$\frac{gH_1}{N_1^2 D_1^2} = \frac{gH_2}{N_2^2 D_2^2} \quad . \quad . \quad . \quad \text{from equation (10.8)}$$

$$\therefore \; H_2 = H_1 \left(\frac{N_2 D_2}{N_1 D_1}\right)^2$$

$$= 4{\cdot}27 \times \left(\frac{3\,185 \times 0{\cdot}4}{1\,800 \times 0{\cdot}2}\right)^2 = 53{\cdot}48 \text{ m}$$

NOTE: The specific speed for the large pump is given by

$$N_s = \frac{3\,185 \times 0{\cdot}92^{1/2}}{53{\cdot}48^{3/4}} = \underline{154{\cdot}5 \text{ rev/min}}$$

This agrees with that for the small pump.

From Fig. 10.10, a suitable type of pump would be a mixed flow design.

3. A quarter-scale turbine model is tested under a head of 11 m. The full-scale turbine is to use a head of 30 m and to run at 430 rev/min. At what speed should the model test be made?

If, during the test, the model produces 100 kW at the point of maximum efficiency, using a flow of 1·08 m³/s of water. What will be the power output of the full-scale turbine and the flow rate required at the point of maximum efficiency?

(*Ans.*: 1042 rev/min; 7·196 kW; 28·52 m³/s)

4. A water turbine is to develop 5 MW at 120 rev/min using a head of 32 m. A scale model test conducted at 210 rev/min with a head of 3·6 m uses a flow rate of 0·22 m³/s at the point of maximum efficiency.

Determine the specific speed and efficiency of the machines, the flow rate for the turbine and the power output of the model.

What type of turbine is appropriate to this performance?

(*Ans.*: 111·5 rev/min; 89·2%; 17·85 m³/s; 6·93 kW; Francis)

5. A hydro-electric power station is to use turbines which have a specific speed of 188 rev/min at 82% efficiency. The head and quantity of water available are 60 m and 33 m³/s respectively. The generator design requires the turbines to run at 480 rev/min. Determine the number of turbines to be installed and the power output of the station. What type of turbine would be required?

(*Ans.*: 4; 15·93 MW; Francis)

6. A pump running at 1400 rev/min delivers 0·03 m³/s at the point of maximum efficiency against a head of 33·5 m. A geometrically similar pump has linear dimensions 50% greater and operates at 1200 rev/min. Determine the head produced by the larger pump at the point of maximum efficiency and the discharge at this point.

What is the specific speed of this range of pumps and the relative power consumption of the two pumps? (*Ans.*: 55·37 m; 0·087 m³/s; 17·41 rev/min; 4·78)

7. The efficiency, η, of a rotodynamic pump depends on the speed, N, the impeller diameter, D, the fluid density, ρ, the flow rate, Q, the head, expressed as gH and the power, P. Show that

$$\eta = \phi\left[\frac{P}{\rho N^3 D^5};\ \frac{Q}{ND^3};\ \frac{gH}{N^2D^2}\right]$$

A pump supplying 0·3 m³/s of water develops a head of 20 m at 2000 rev/min and requires a power supply of 100 kW. For dynamically similar conditions at the point of maximum efficiency, what are the rotational speed, the head and flow rate when the power supply is reduced by 30%?

(*Ans.*: 1776 rev/min; 15·77 m; 0·27 m³/s)

8. The power required to drive a pump at the design point is 6 kW at 1465 rev/min. The discharge is then 25 kg/s against a head of 15·25 m.

Determine the size ratio of another similar pump which would operate at the point of maximum efficiency from a motor which produces 4·7 kW at 2500 rev/min. What would be the head and discharge from this pump under these conditions?

(*Ans.*: 0·691; 21·2 m; 14·1 kg/s)

9. A pump for use in an irrigation scheme is to raise 4500 m³/h of water against a head of 2·5 m with the pump working at maximum efficiency.

Tests on a geometrically similar pump of smaller size gave the following results at a constant speed of 1000 rev/min.

Quantity (m³/s)	0·045	0·057	0·062	0·068	0·076	0·079	0·083	0·091
Head (m)	14·0	13·7	13·4	12·95	12·2	11·73	11·0	8·84
Power (kW)	10·07	11·41	11·86	12·31	12·76	12·98	13·20	13·50

Determine the specific speed of the series and the speed at which the irrigation pump should run. What type of pump should be used?

(*Ans.*: 42·23 rev/min; 75·1 rev/min; centrifugal)

11 Centrifugal pumps

11.1 Introduction The centrifugal pump is one of the most common types of turbomachine and the theory is now analysed in detail. A similar analysis could, however, be applied to axial flow pumps, turbines, fans, etc., with appropriate essential differences.

In Chapter 5, the momentum equation is used to determine the force on a curved blade and this forms the basis of the action of a turbomachine. It is common to resolve the absolute velocity of the water into components in the radial, tangential and axial direction; the force in any direction then derives from the change in momentum in that direction.

The velocity component tangential to the rotor circumference is called the *velocity of whirl*; the force in the whirl direction is that which gives rise to the torque and power and is therefore the most important force considered in turbomachine analysis.

11.2 The Euler equation In this analysis, the following notation is used:

Absolute velocity at inlet	V_1	
Absolute velocity at outlet	V_2	
Relative velocity at inlet	u_1	
Relative velocity at outlet	u_2	
Tangential velocity of blade tip at inlet	v_1	$(= \omega r_1)$
Tangential velocity of blade tip at outlet	v_2	$(= \omega r_2)$
Velocity of whirl at inlet	v_{w_1}	
Velocity of whirl at outlet	v_{w_2}	
Radial velocity at inlet	v_{r_1}	
Radial velocity at outlet	v_{r_2}	

Fig. 11.1 shows the relative velocity triangles at inlet and outlet of a typical blade of a radial flow (outward) pump impeller. The directions of u_1 and u_2 are tangential to the blade tips for smooth flow; the directions of v_1 and v_2 are tangential to the impeller circumference at the blade tips.

The torque about the shaft axis, exerted on the water, is given by the rate of change of the angular momentum of the water about that axis.

Momentum of water at inlet/sec = mass of water/sec × tangential velocity

$$= \rho Q \times v_{w_1}$$

∴ moment of momentum (or angular momentum) of water/sec

$$= \rho Q v_{w_1} r_1$$

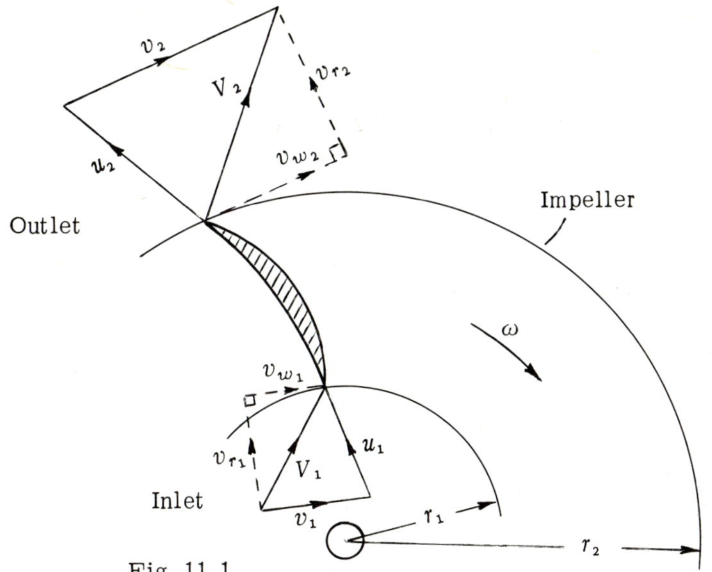

Fig. 11.1

Similarly angular momentum of water/sec at outlet

$$= \rho Q v_{w_2} r_2$$

$$\therefore \text{ torque } = \text{ change of angular momentum/sec}$$

i.e.

$$T = \rho Q (v_{w_2} r_2 - v_{w_1} r_1) \tag{11.1}$$

This is known as the *Euler equation.*

The power input, $P = T\omega$

$$= \rho Q (v_{w_2} r_2 - v_{w_1} r_1)\omega$$

But $\omega r_1 = v_1$ and $\omega r_2 = v_2$, so that

$$P = \rho Q (v_{w_2} v_2 - v_{w_1} v_1) \tag{11.2}$$

Equation (11.2) represents the energy input per second but in pump analysis, it is often more convenient to work in terms of the head produced, rather than the power. The head H represents the energy per unit weight (see Art. 4.2) and so

$$P = \rho g Q H$$

Thus

$$H = \frac{\rho Q (v_{w_2} v_2 - v_{w_1} v_1)}{\rho Q g}$$

$$= \frac{v_{w_2} v_2 - v_{w_1} v_1}{g} \tag{11.3}$$

Equation (11.3) represents the *Euler head.*

11.3 Application to centrifugal pumps The general arrangement of a centrifugal pump is shown in Fig. 10.6. Water enters the casing in an axial direction but is turned through 90° before entering the impeller. It then passes radially through the blades to enter the volute chamber.

At inlet to the impeller, the water velocity has no component in the tangential direction, i.e. the velocity of whirl v_{w_1} is zero and the velocity trangle is right-angled. Hence equations (11.2) and (11.3) reduce to

$$P = \rho Q v_{w_2} v_2 \qquad (11.4)$$

and
$$H = \frac{v_{w_2} v_2}{g} \qquad (11.5)$$

The relative velocity triangles for this case are shown in Fig. 11.2.

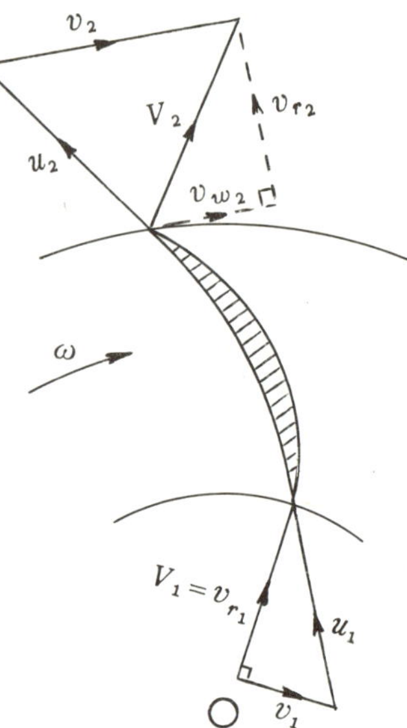

Fig. 11.2

The Euler head is greater than the actual head produced, due to losses. The actual head produced comprises the suction head H_s, the discharge head H_d, the friction losses in the suction and delivery pipes, H_{f_s} and H_{f_d} and the discharge velocity head $\frac{v_d^2}{2g}$. This head is known as the *manometric head*, H_m, since it is the head which would be registered by a manometer connected across the pump at inlet and outlet if the inlet and outlet pipes were the same diameter.

Thus
$$H_m = H_s + H_d + H_{f_s} + H_{f_d} + \frac{v_d^2}{2g}$$

The distribution of these heads is shown in Fig. 11.3.

The ratio of the manometric head to the Euler head is called the manometric efficiency,

i.e.
$$\eta_m = \frac{H_m}{\dfrac{v_{w_2} v_2}{g}} = \frac{gH_m}{v_{w_2} v_2} \qquad (11.6)$$

The difference between $\frac{v_{w_2} v_2}{g}$ and H_m is made up of hydraulic losses such as friction, bends, eddies, etc., and the effectiveness of the volute casing or diffuser in converting the impeller exit velocity to pressure energy. In addition to these hydraulic losses, there are also mechanical losses such as bearing friction and the overall efficiency of the pump will be less than the manometric efficiency.

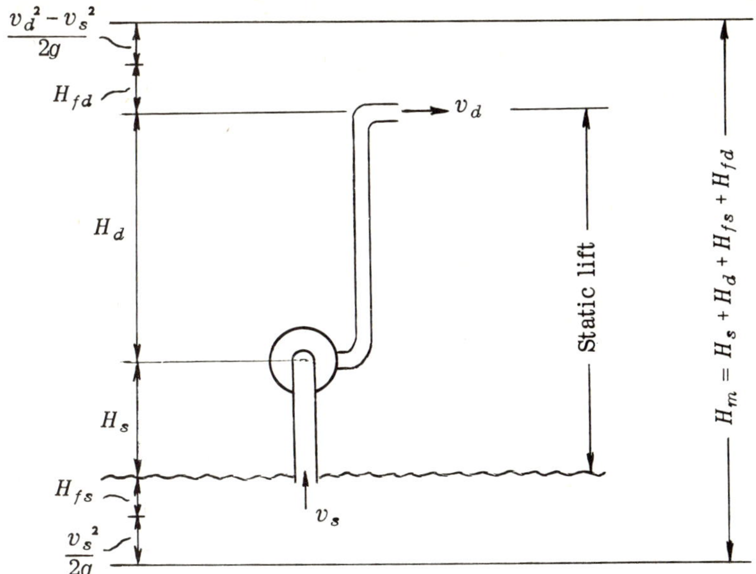

Fig. 11.3

11.4 Forward facing, radial and backward facing impeller blades Fig. 11.2 shows a pump impeller with backward facing blades but there are also designs with radial and forward facing blade directions at exit. Fig. 11.4 shows these arrangements, from which it will be seen that, for the same values of u_2 and v_2, v_{w_2} (and hence the Euler head) is greatest with forward facing blades and least with backward facing blades.

In Fig. 11.4, θ represents the outlet angle of the water relative to the whirl direction. Then, in all cases,

$$v_{w_2} = v_2 - v_{r_2} \cot\theta \quad ,$$

cot θ being negative when θ is greater than 90°.

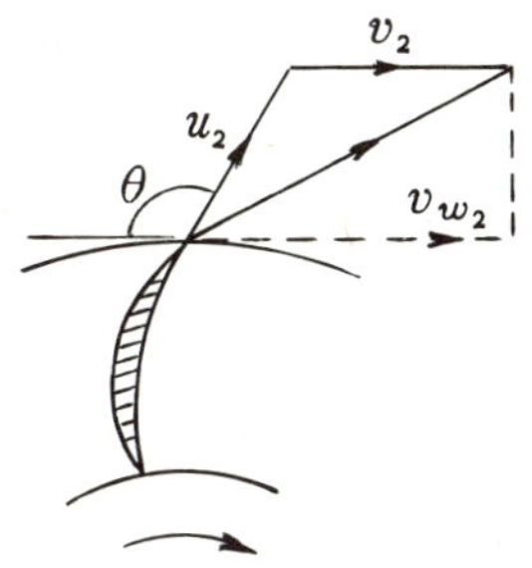

Forward facing blades

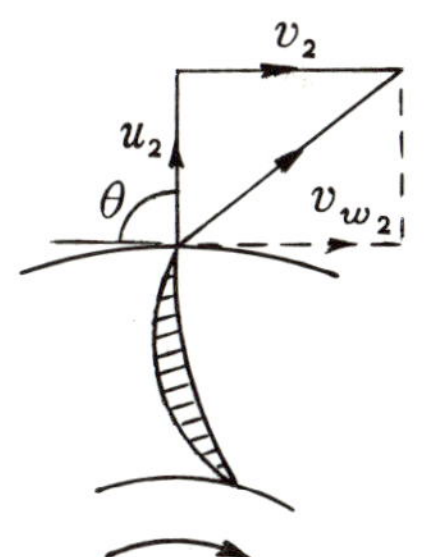

Radial blades

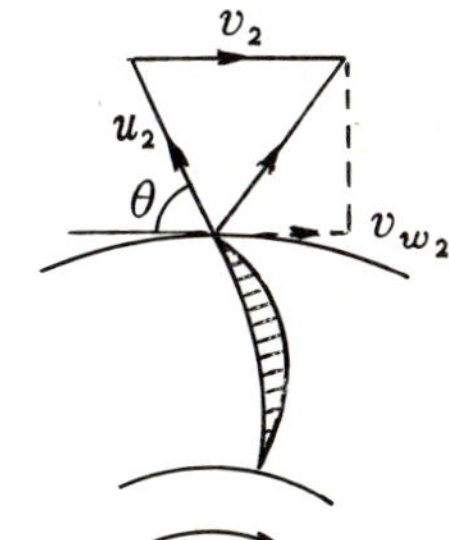

Backward facing blades

Fig. 11.4

Thus the Euler head,

$$\frac{v_{w_2} v_2}{g} = \frac{v_2}{g}(v_2 - v_{r_2} \cot\theta)$$

If the outlet flow area is A_2, then $v_{r_2} = \frac{Q}{A_2}$, which is a constant.
Also $v_2 = \omega r_2$, so that

$$\text{Euler head} = A\omega^2 - B\omega Q \cot\theta \qquad (11.7)$$

where A and B are constants.

This equation is represented graphically in Fig. 11.5 for forward facing, radial and backward facing blades at a constant speed ω, with allowance for the sign of $\cot\theta$.

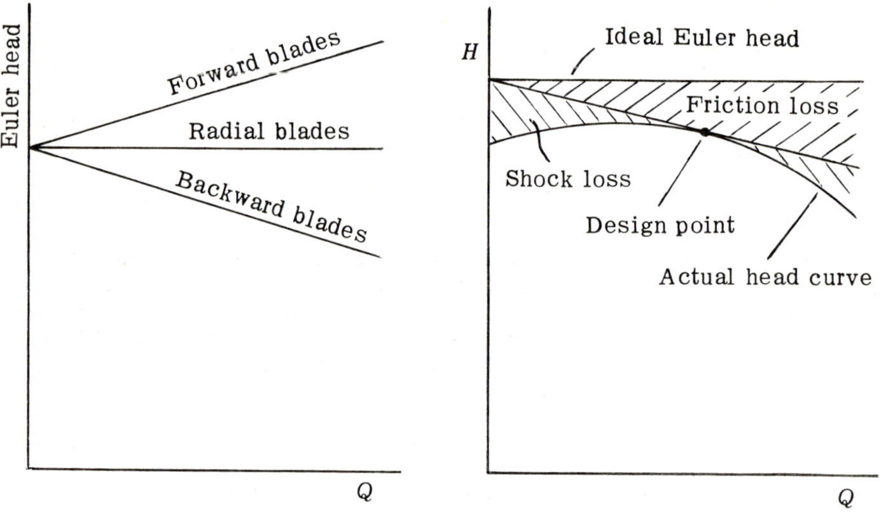

Fig. 11.5 Fig. 11.6

The actual head – flow rate curves will be modified by
(*a*) friction effects, which increase as Q increases, and
(*b*) shock losses away from the design point, when the liquid flow is not at the correct inlet angle.

These effects are shown in Fig. 11.6, for the case of radial blades.

The actual head – flow rate curves for the three types of blade are shown in Fig. 11.7. It will be seen that forward-facing blades give a greater head for a given flow rate and it would therefore be expected that pumps would generally be of this design. However, this is not so – the *slopes* of these curves are of great importance.

Where the curve has a negative slope at all points, such as with backward facing blades, there is only one flow rate for any value of H and the flow is stable but for a curve having positive and negative slopes, there are two values of Q for some values of H. This can lead to unstable operation, especially when pumps are used in parallel and so pumps with backward facing blades are preferred.

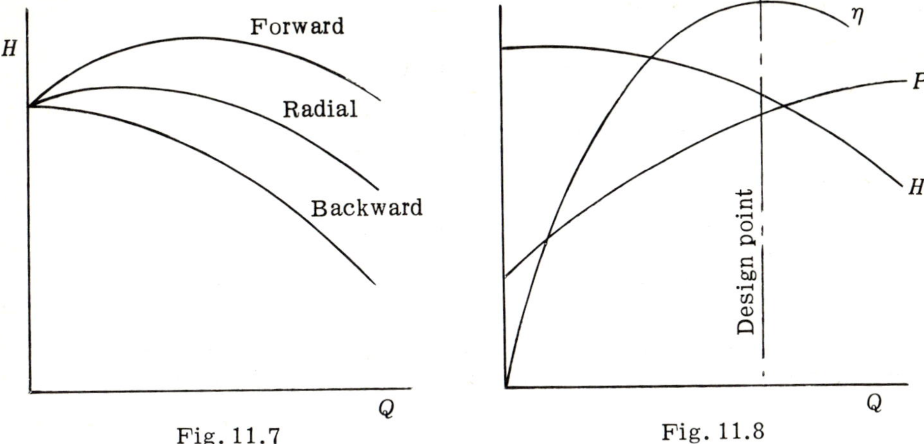

Fig. 11.7

Fig. 11.8

11.5 Performance characteristics The parameters of turbomachine performance, developed in Art. 10.2, are head, power and efficiency, each being a function of the flow rate, Q. Fig. 11.8 shows the relations of these quantities for a centrifugal pump with backward facing blades, running at constant speed. The design point is that corresponding with maximum efficiency; it is at this point that the specific speed is evaluated, in order to assist in the selection of a pump for a particular application.

11.6 System characteristics and matching A pump is installed in a system to give a static lift, h, between suction and delivery points but, in addition, it will also have to overcome losses due to friction, bends, valves, etc. These losses are all proportional to V^2 and hence proportional to Q^2 for a given cross-sectional area of flow. Thus a *system characteristic* at constant speed may be expressed in the form

$$H = h + kQ^2 \tag{11.8}$$

where H is the required output head and k is a constant.

This relation is shown in Fig. 11.9.

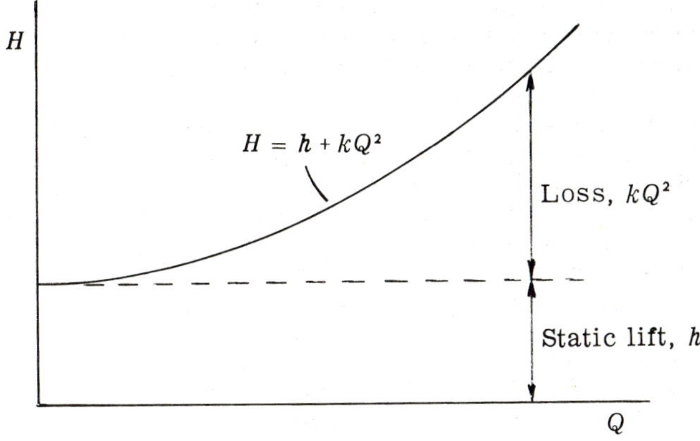

Fig. 11.9

Having selected a pump from specific speed considerations, the actual running point is determined from the combination of the pump and system characteristics; this is known as *matching* and is illustrated in Fig. 11.10. The matching point is at the intersection of the two characteristic curves which should ideally correspond with the maximum efficiency point for the pump.

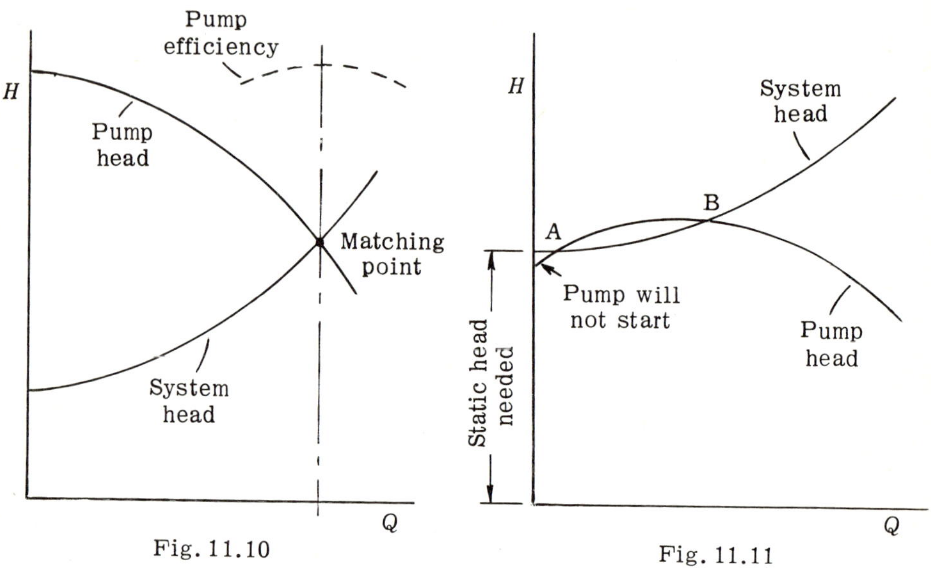

Fig. 11.10 Fig. 11.11

In the case of forward facing blades, the pump characteristic initially rises with increase of flow and it is possible that the system characteristic may intersect it at two points, Fig. 11.11. Thus there will be two operating points, A and B, and variations in the speed of the driving motor or in the delivery head may cause the discharge to oscillate between these positions, leading to unstable running.

This possibility is greatly increased when two unstable pumps are operating in parallel, discharging into a common delivery pipe; a change in the operating point in one pump may cause the second pump to start oscillating, which in turn will affect the running of the first pump. It is therefore necessary that pumps operating in parallel should have stable characteristics.

A further problem can arise if the system head is greater than the pump head at zero discharge. The pump will then be unable to start unless the discharge pipe is partly emptied to reduce the opposing head.

1. *A centrifugal pump is required to deliver 0·75 m^3/s of water against a manometric head of 16 m. At outlet, the relative velocity of the water is inclined at 120° to the tangential velocity of the rotor (i.e. backward facing blades) and the radial velocity is 0·27 times the rotor tangential velocity. The manometric efficiency is 85% and the width of the impeller at exit is 0·1 times the outside diameter.*

Determine the outside diameter of the impeller and the speed of rotation.

From equation (11.6), $$\eta_m = \frac{gH_m}{v_{w_2} v_2}$$

$$\therefore \; v_{w_2} = \frac{9{\cdot}81 \times 16}{0{\cdot}85 v_2} = \frac{184{\cdot}66}{v_2} \qquad (1)$$

Fig. 11.12(*a*) shows the relative velocity diagram at exit.

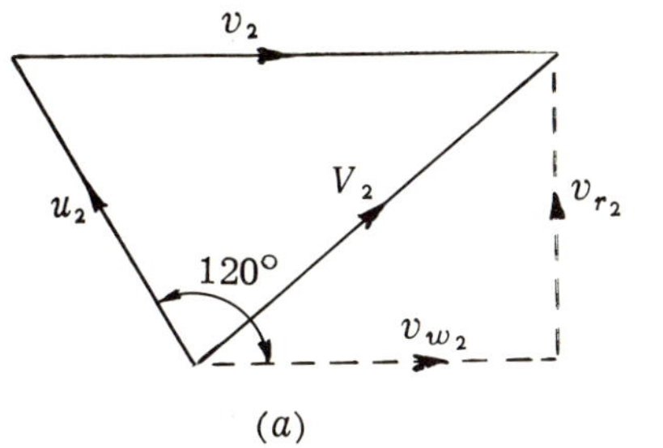

(*a*)

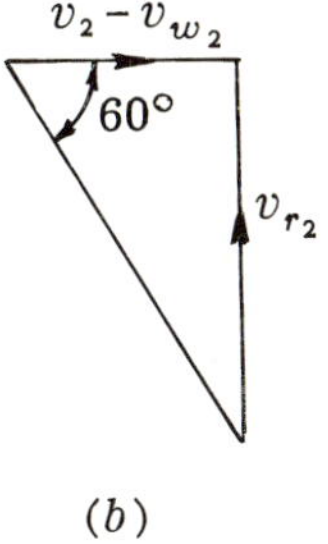

(*b*)

Fig. 11.12

From Fig. 11.12(*b*),

$$\tan 60^\circ = \frac{v_{r_2}}{v_2 - v_{w_2}}$$

But $v_{r_2} = 0{\cdot}27 v_2$,

$$\therefore \; 1{\cdot}732 = \frac{0{\cdot}27 v_2}{v_2 - v_{w_2}}$$

from which $$v_{w_2} = 0{\cdot}844 v_2 \qquad (2)$$

Hence, from equations (1) and (2),

$$v_2 = 14{\cdot}79 \text{ m/s}$$

But $$v_2 = \frac{\pi D N}{60}$$

from which $$DN = \frac{60 \times 14{\cdot}79}{\pi} = 282{\cdot}47 \qquad (3)$$

Quantity flowing, $$Q = \text{exit area} \times v_{r_2}$$

i.e. $$0{\cdot}75 = (\pi D \times 0{\cdot}1D) \times (0{\cdot}27 \times 14{\cdot}79)$$

from which $$D = \underline{0{\cdot}773 \text{ m}}$$

From equation (3), $$N = \frac{282{\cdot}47}{0{\cdot}773}$$

$$= \underline{365{\cdot}4 \text{ rev/min}}$$

2. *A centrifugal pump running at 1 000 rev/min gave the following results in a test:*

Discharge (m^3/s)	*0*	*0·075*	*0·151*	*0·227*	*0·302*	*0·359*
Head (m)	*23·0*	*22·5*	*21·8*	*19·8*	*14·3*	*0*

(a) The pump is to be connected to 300 mm diameter suction and delivery pipes with the discharge point 13 m above sump level. The total length of the suction and delivery pipes is 75 m and the entry loss is equivalent to a further 7 m of pipe. If the friction coefficient for the pipes is 0·006, calculate the discharge from the pump.

(b) If the speed is reduced to 800 rev/min, determine the new flow rate, assuming the efficiency to be unchanged.

(*a*) Pipe velocity, $$V = \frac{Q}{\frac{\pi}{4} \times 0{\cdot}3^2} = 14{\cdot}147Q$$

Effective length of pipe $= 75 + 7 = 82$ m

$$\therefore \text{ friction loss in pipe} = \frac{4flV^2}{2gd}$$

$$= \frac{4 \times 0{\cdot}006 \times 82\,(14{\cdot}147Q)^2}{2 \times 9{\cdot}81 \times 0{\cdot}3}$$

$$= 66{\cdot}92Q^2$$

$$\text{Exit loss} = \frac{V^2}{2g} = \frac{(14{\cdot}147Q)^2}{2 \times 9{\cdot}81} = 10{\cdot}2Q^2$$

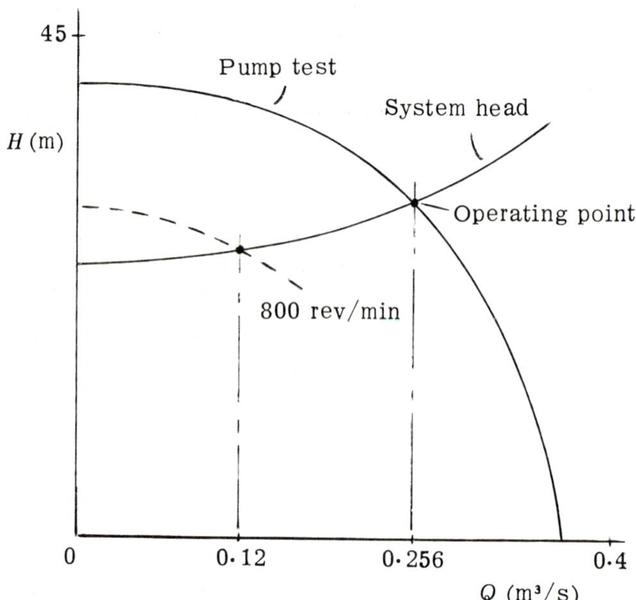

Fig. 11.13

System characteristic, H = static lift + losses

$$= 13 + (66{\cdot}92 + 10{\cdot}2)Q^2$$

$$= 13 + 77{\cdot}12Q^2$$

For the values of Q used in the test results, corresponding values of H are as follows:

H (m) :— 13 13·44 14·76 16·97 20·03 22·94

The graphs of pump head and system head against discharge are shown in Fig. 11.13, from which the intersection gives a discharge of 0·256 m³/s.

(*b*) At a given efficiency, $\frac{Q}{ND^3}$ is a constant

$$\therefore \quad Q_2 = Q_1 \times \frac{N_2}{N_1} \qquad \text{since } D \text{ is unchanged}$$

Also $\frac{gH}{N^2D^2}$ is a constant,

$$\therefore \quad H_2 = H_1 \times \frac{N_2^2}{N_1^2}$$

The pump characteristic at 800 rev/min can now be obtained by multiplying the original discharge values by 0·8 and the corresponding head values by $0{\cdot}8^2$.

The results are as follows:

Discharge (m³/s)	0	0·06	0·121	0·182	0·242	0·287
Head (m)	14·7	14·4	14·0	12·7	9·2	0

This curve is plotted on the original diagram and intersects the system head curve, which remains unchanged, at a discharge of 0·12 m³/s

3. A centrifugal pump delivers 0·012 m³/s of water when the pressure difference across the pump is 0·33 MN/m² and the speed is 2 100 rev/min. The manometric efficiency is 0·83. If the impeller is 0·2 m in diameter and the exit width is 0·012 m, determine the blade exit angle. (*Ans*.: 21·75°)

4. A centrifugal pump is to be used in a system having a head characteristic given by $H = (12 + 7\,410Q^2)$ m where Q is the discharge in m³/s. In a trial on the pump, the following results were obtained:

Speed (rev/min)	1 215	1 212	1 210	1 202	1 200	1 195	1 190
Head (m)	25·6	27·0	26·8	26	24·3	20	13·1
Discharge (m³/s)	0	0·011	0·018	0·025	0·032	0·041	0·049

Correct the table to a speed of 1 200 rev/min and determine the pump discharge at 1 200 rev/min with the system. (*Ans*.: 0·036 5 m³/s)

5. In a test on a centrifugal pump the following results were obtained:

Speed (rev/min)	2 160	2 140	2 130	2 120	2 105	2 120
Head (m)	34·44	37·19	34·14	29·41	21·34	14·33
Discharge (m^3/s)	0	0·007 1	0·011 8	0·016 7	0·023 2	0·028 6
Power (kW)	–	4·77	5·80	6·56	7·16	7·16

Plot curves of head and efficiency against quantity flowing for a speed of 2 100 rev/min and hence determine the power requirement at 2 100 rev/min when used in a system with a head demand, $H = (13{\cdot}4 + 59\,300Q^2)$ m where Q is the quantity flowing in m^3/s. (*Ans.*: 6·35 kW)

6. A centrifugal pump is required to discharge 0·57 m^3/s of water with a head of 13 m when rotating at 750 rev/min. The manometric efficiency is 0·8 and the radial velocity is 3 m/s. The loss of head in the pump is $0{\cdot}026\,2V^2$ where V is the impeller absolute exit velocity in m/s.

Determine the impeller diameter, exit width and blade exit angle. (*Ans.*: 0·375 m; 0·16 m; 35·7°)

7. A centrifugal pump delivers 0·22 m^3/s with a head of 25 m at 1 500 rev/min. The manometric efficiency is 75% and the loss of head in the system is $0{\cdot}033V^2$ where V is the absolute velocity at impeller exit in m/s. The exit area of the pump is $1{\cdot}2D^2$ where D is the impeller diameter. Determine the blade exit angle and the impeller diameter. (*Ans.*: 26·3°; 0·266 m)

8. A centrifugal pump impeller has an external diameter of 0·3 m and an outlet area of 0·11 m^2. The backward facing blades make an angle of 145° to the whirl direction. Pressure gauges at the same level above the sump in the suction and delivery lines close to the pump read heads of water of 3·7 m below atmospheric pressure and 21 m above atmospheric pressure respectively when the pump delivers 0·204 m^3/s of water at 1 200 rev/min. The power input is 71·6 kW and pipe friction losses are 3 m of water.

Determine the manometric efficiency and the overall efficiency of the pump. (*Ans.*: 79·4%; 61%)

9. A centrifugal pump delivers 0·94 m^3/s of water against a head of 25 m of water. The outer diameter of the impeller is 1 m and the inner diameter is 0·5 m. The pump speed is 600 rev/min. The radial flow velocity is constant at 6 m/s and the pump efficiency is 0·75. Find the inlet and outlet impeller blade angles relative to the whirl direction, the inlet and outlet blade widths and the power input to the pump if the mechanical efficiency is 0·9. (*Ans.*: 21°; 15·9°; 0·1 m; 0·05 m; 342 kW)

10. The results of a trial on a centrifugal pump are as follows:

Discharge (m^3/s)	0	0·255	0·481	0·750	0·990
Head (m)	132	121	106	78	31

Two such pumps deliver water into a pipeline 0·6 m diameter and 1 000 m long against a static head of 50 m. The friction coefficient in the pipe is 0·005. Neglecting velocity head, determine the discharge if the pumps are coupled (*a*) in parallel, and (*b*) in series. (*Ans.*: 1·35 m^3/s; 0·98 m^3/s)